COMB HONEY PRODUCTION

ROGER A. MORSE
Professor of Apiculture
Cornell University □ Ithaca, New York

WICWAS PRESS
425 Hanshaw Road   □   Ithaca, N.Y. 14850   □   1978

© Copyright 1978 by Roger A. Morse
All rights reserved. Original printed in the U.S.A.
Reprint edition printed in various global facilities; U.S.A. customers receive books printed in the U.S.A.

No part of this publication may be reproduced or transmitted in any form or by any other means, electronic or mechanical, including photocopy, recording, or any information storage or retrieval system now known or to be invented, without prior written permission from the publisher, except by a reviewer who wishes to quote brief passages in connection with a review written for inclusion in a magazine, newspaper or broadcast.

Library of Congress Catalog Card Number 78-51041

Reprint Edition
Reprinted with a new cover 2014
Cover design modified from work by Camden Leeds and Jeff Shaw, Seedsgreenprinting.com
Cover photographs © Lawrence John Connor
New material © Wicwas Press, LLC
www.wicwas.com

ISBN 978-1-878075-34-5

Contents

Preface	7
Introduction — What is Comb Honey?	9
Comb honey history	10
Honey adulteration	11
I. Equipment for Comb Honey Production	14
Bottomboards	15
The brood nest	17
Frames	18
Follower boards	20
Covers and innercovers	20
Hivestands and grass	21
The great argument — the best size hive	24
II. Supers and Sections	27
Split versus solid sections	29
Supers	30
The Cobana super and sections	32
Foundation	35
Starters versus full sheets of foundation	36
Odd size cell diameters	36
When to put foundation into sections	37
Mounting boards and methods of fastening foundation	39
Preparation of sections and supers	43
Queen excluders	45

4 CONTENTS

III. The Apiary Site	46
Commercial sites	47
Keeping bees in congested areas	48
Physical requirements of good apiary sites	49
Outbuildings	51
IV. Outline of Seasonal Management for Comb Honey Production	52
March-April — unpacking and the first inspection	53
April-May — the first major inspection	56
May-June — swarm prevention	63
Preparation for the honey flow	67
Swarm control	70
Late summer and fall management	74
September — the food chamber	75
October-November — packing	76
V. A Special Management Technique	77
Reversing supers	78
Swarm prevention	79
Preparation for the following year	80
VI. Cut Comb Honey	81
Management for cut comb versus comb honey	82
Savings in cut comb honey production	83
Preparation of the frames	84
Cleaning the frames	85
Cutting cut comb honey	86
Packaging cut comb honey	86
VII. Supering	88
Supering for comb honey	88
Leveling colonies	90
Turning comb honey supers end for end	92
Closing upper entrances	92
Starter sections	93
Go-backs	94

CONTENTS

VIII. Removing Comb Honey Supers 98
    Smoke particles and the danger of oversmoking 99
    Robbing 100
    Bee escapes 100
    Bee blowers 102
    Repellents 103

IX. Fumigating and Packaging of Comb Honey 105
    Wax moths 106
    Freezing to protect against moths 108
    Fermentation 109
    Cleaning sections 111
    Packaging 112
    Sun bleaching comb honey sections 113
    Second labels (back labels) 114
    Shipping comb honey 115
    Weight laws 116

X. Showing and Grading Comb Honey 117
    Official grades for comb honey 117
    Judging comb honey 120

Index 127

Preface

Rearing queen bees and producing comb honey are arts; they are the two most difficult and specialized aspects of beekeeping. Persons who can successfully do either must be knowledgeable in the biology of the honey bee. Queen rearing continues to be widely practiced because of the need for young queens by northern honey producers. Interest in comb honey production has waned because of increased production costs, a public not too knowledgeable about honey and the ease with which liquid honey may be produced.

I know of no beekeeper who makes a full-time living producing comb honey in the U.S. today. As late as 1950 three men did so in New York State. At the turn of the century millions of sections of comb honey were produced across the country. This is another indication of trends in modern agriculture. Still, a few commercial beekeepers and many hobbiests prefer to produce comb honey. This book has been prepared for those persons.

Comb honey production is a challenge for all beekeepers. No natural product is so delicious, tempting or difficult to produce. Any beekeeper may occasionally produce a good, or even a perfect section of comb honey; however, to do so year after year, season after season, is not an easy task. It can be done only by persons thoroughly familiar with bee behavior.

This book is only a summary of what we know and the way in which this knowledge has practical application. As a result of research conducted during the past fifteen years, we have a better understanding of why bees react as they do and the factors which dictate their behavior. Still, the methods recommended in this text date from the late 1800's. Since that time our agriculture has changed and the beekeeper must be aware of the potentials and limitations for honey production in his area. Without a good source of nectar producing plants even the best beekeeper will fail.

The most important thing in beekeeping is to have bees where nectar producing plants abound. The second factor is colony management. Control of bee diseases is third in importance. Far behind these are questions of hive size, races of bees, control of pests, etc.

Introduction—
What is Comb honey?

Comb honey is honey in new, natural comb built by bees and fastened in a small, light-weight wooden or plastic frame. It is sometimes called section honey or card honey. Pieces of honey in the comb cut from large frames are usually referred to as cut comb honey or chunk honey. Honey is a natural product. Whether it be in the comb, or liquid honey which has been removed from the comb, it contains no additives.

Methods for making comb honey are only a little over 100 years old. Still, literature 200 to 2000 years old makes reference to virgin honey, meaning honey in new, white comb. Virgin honey was always a rare commodity both because ancient beekeepers did not have good management schemes and because they usually had to kill the colony to harvest the honey. The comb honey producer seeks to produce the same virgin honey but in quantity and in a package which is convenient for man to handle.

When eating comb honey one may chew and consume the wax. New, white beeswax is nearly flavorless. Old black comb has an objectionable flavor. Honey bees collect plant resins and gums and use these to strengthen their combs. These plant products, as well as pollen, give the old comb its color and poor flavor. Fortunately, liquid honey stored in old comb does not pick up these off-flavors and when extracted (separated) from the comb has a good flavor.

Beekeepers agree that comb honey has a better flavor than liquid honey, even when the two are produced from the same plant nectar. When liquid honey is extracted it is exposed to a great deal of air and some of its delicate flavor is lost. Liquid honey which is sold in supermarkets is usually heated to destroy yeasts and to prevent fermentation. When this is done carefully it does little harm to the delicate honey flavor; still, comb honey, because it is not subjected to these processes, has the better flavor.

Comb honey history

According to Frank Pellett, who wrote *History of American Beekeeping* in 1938, the credit for inventing the comb honey section belongs to J. S. Harbison of California. Harbison said he perfected the method on Christmas Day in 1857. He called his product "section honey-box".

The early sections were made using various woods but it was finally decided that wood from the basswood tree made the finest sections and would not impart a flavor to the honey. That basswood lumber should be used to make comb honey sections is interesting for the basswood tree produces a nectar with a resulting honey considered by some, including the author, to be of finest quality.

The early sections were made using four pieces of wood nailed together. Later it was found that basswood was suitable for a one-piece section which could be folded into a box or section. A variety of comb honey section box sizes and designs have been developed; today only two are manufactured and widely used in the United States.

Honey adulteration

While the first comb honey sections marketed met with public acceptance because of their attractiveness and the flavor of the product, there was still another reason why section comb honey was popular with beekeepers in the early days of the industry. Most of the consuming public was then, and is now, aware of the fact that comb honey is a natural product and not adulterated. In the 1800's men learned how to make a sugar syrup by converting the starch in corn into the sugar glucose. Since corn was available in great quantities the resulting product was cheap and easy to produce. About the time the commercial beekeeping industry started to be successful in the 1870's and 1880's, much of the honey and maple syrup on the market was adulterated with glucose. The Nation's beekeepers were among the very first people to press the federal government for enactment of the Pure Food and Drug Laws. One of the valued possessions in the Cornell University Library is a copy of a petition requesting enactment of such laws which went to Congress from our Nation's beekeepers in 1880.

The first Pure Food and Drug Laws were passed in the United States in 1906. Final passage of the Pure Food and Drug Laws may very well have been delayed by the beekeepers who were greatly irritated by the famous "Wiley Lie". Wiley, a chemist with the USDA, wrote a popular article in 1881 saying that comb honey was made with paraffin and the cells were filled with glucose. This prompted A. I. Root, founder of the A. I. Root Co., to offer a one-thousand-dollar reward for anyone who could make a comb honey section which would be indistinguishable from that made by honey bees; the reward, naturally, was never collected. While there are no precise figures available, it is possible that between a third and a half of the honey produced in the United States at the turn of the century was comb honey.

12 COMB HONEY PRODUCTION

Since it is cheaper and easier to produce liquid honey, the decline in the production of comb honey started soon after the passage of the Pure Food and Drug Laws; it was speeded up by World War I, during which time there was a sugar shortage and the price of honey increased greatly.

Between 1906 and 1971 there were almost no cases of honey adulteration in the United States. Soon after the price of honey began to rise in 1971, there appeared on the market a very small number of adulterated and imitation honeys. This unfortunate circumstance is being watched closely by the industry which hopes to retain the confidence of the American public in pure honey. Honey is a luxury product. It is costly and time consuming to produce. Comb honey is even more so. So long as there is a beekeeping industry, people will no doubt need to be concerned with substitutes, imitations and adulteration.

In 1976 researchers in the U.S. Department of Agriculture found a method of detecting honey adulteration when high fructose corn syrup was used as an adulterant. Adulteration of honey with other sugars is not too difficult to determine but it had been feared the chemical nature of the newly developed high fructose corn syrup was such that it would be difficult to ascertain when this substance was used. It was determined by researchers that the formation of sugar through photosynthesis is different in different groups of plants. High fructose corn syrup fructose is different from the fructose found in honey; the carbon isotope ratios are not the same. The fact that this is true gives science a new tool with which to unmask those who would adulterate and make false claims for their products. This research has been followed closely by the Honey Industry Council and state and federal researchers in apiculture. The cost of analysis is not too high. The Council may bear the expense of such analysis where there is good reason to suspect a problem.

INTRODUCTION—WHAT IS COMB HONEY?

It is known that one large shipment of imported honey in the early 1970's was adulterated. However, the beekeeping industry has pursued this matter with vigor and this, in itself, has probably served to deter those who would cheat. The author is not aware of any other cases of adulteration in recent years and has good reason to state firmly that the liquid honey on the market today, at least in this country, is free of adulterants.

I.
Equipment for Comb Honey Production

In producing comb honey we are concerned that neither the wood in the sections nor the comb surface be travel stained. Travel stain is the discoloration of the comb or wood caused by bees walking over it and accidentally dropping bits of pollen and propolis which color the wax. Cappings on comb honey sections should be as white and clean. In honey shows most judges will disqualify badly travel stained sections and penalize those with even light travel stain.

It is also important to reduce the deposition of propolis which could likewise stain the wood. Preventing travel stain and deposition of propolis is in part a question of management; however, selecting and using the proper equipment can help in this regard.

Old time comb honey producers selected bees which would use little propolis. They were very much concerned that the wax surface of their comb honey sections be as white as possible. Since most beekeepers today produce liquid honey only, the selection of bees for comb honey production alone is scarcely practical. Thus, even greater weight must be placed on management and the proper equipment for comb honey production.

EQUIPMENT FOR COMB HONEY PRODUCTION   15

Bottomboards

The late Dr. C. C. Miller of Illinois was one of the world's most famous comb honey producers. He used and preferred a bottom board which is two inches deep. Carl Killion, retired state apiary inspector from Illinois, and also a famous comb honey producer, did the same. Miller described his bottomboard as a box with three sides and without a top. Both Miller and Killion used and recommended a rack which fitted inside the bottomboard so as to deter comb building down into the large, open space. George Demuth, who wrote extensively about comb honey production and who was with the USDA for a number of years before becoming editor of *Gleanings in Bee Culture,* while he did not advocate a deep bottomboard in his commercial comb honey production bulletin, did stress in his papers on swarm control that proper ventilation would deter swarming. While congestion is the primary cause of swarming there are factors which will tend to encourage or discourage it. Miller's deep bottomboard gives the bees a clustering space and adequate room for ventilation. Ventilation is also important when bees are removing moisture from unripe honey.

Many comb honey producers do not use deep bottomboards and they are never used in extracted honey production. As a result of watching colonies on warm summer evenings when there is a good honey flow, I am of the opinion that the deep bottomboard is helpful. Miller left his deep bottomboard, with its rack, in place all year and thought it was helpful insofar as wintering was concerned. (Miller wintered his bees in a cellar.)

The precise design of the bottomboard is not especially important so long as it is of a reasonable depth. Many beekeepers have used plywood successfully; others have found that old boards, not satisfactory for making hive bodies or hive parts, will work quite well. Because of their close prox-

1. A Miller-type bottom board with rack. The sides of the bottom board are about two inches deep. The rack, which is built so as to have a proper bee space between it and the bottom bars of the frames in the super above, deters comb building into the large space in the bottom board. Only rarely will bees build comb below the rack.

imity to the ground, bottom boards tend to rot and decay rapidly. Bottomboards are the only pieces of hive equipment which should be routinely treated with wood preservative. (Hive stands, where they are used, should also be treated with a wood preservative.) Normally, new bottomboards are soaked in a solution of kerosene and pentachlorophenol for 24 hours. During this period of time the pine wood will usually absorb sufficient preservative that it will be protected and have a life of 25 to 30 years. After soaking, the bottomboards should be exposed to the weather for several months before they are used. During this time the kerosene evaporates while the wood preservative remains in place. Bottomboards treated with preservative are usually not painted until they have been in use for a year or more. Pentachlorophenol is not toxic to honey bees but fresh kerosene fumes can cause difficulty. It is for this latter reason that treated bottomboards should be thoroughly aired and dried before being used. The rack which is placed in the bottomboard need not be treated with wood preservative.

The brood nest

A standard full depth frame[1] contains about 6,800 cells. A queen lays at a rate of 1,000 to 1,500 eggs per day as a maximum. Worker bees emerge from their cells 21 days after the eggs are laid and drones 24 days. In theory, a queen has all the room she needs to lay in nine frames in a standard 10-frame super.

As noted under the management section below, colonies of

---

[1] The outside dimensions of a standard frame are 9 1/8 by 17 5/8 inches; for the specifications for other hive parts a one-page circular entitled "Plans and dimensions for a 10-frame bee hive" may be obtained by writing the Bio-environmental Bee Laboratory, Agricultural Research Service, USDA, Beltsville, Maryland 20705.

18   COMB HONEY PRODUCTION

honey bees which are used for comb production are allowed to grow into three or four supers in the spring. They are not reduced to a single super brood nest until the first or second day of the honey flow. Since bees will move propolis from one area to another in a hive, it is important that the single super and frames which are used for the final brood nest contain a minimum of propolis. It is usually best to have a supply of scraped supers ready when the colonies are reduced to one super; as frames are selected for the final brood nest they are placed in a clean super.

A few beekeepers, and only those who have a surplus of time on their hands, scrape the top bars of the frames and coat them with paraffin. Bees collect propolis only when it is needed, and if the top bars are paraffined the bees are less likely to collect propolis and varnish the top bars with it. The chief concern, of course, is that if propolis is present in any quantity that it will travel stain the sections immediately above the brood nest.

Frames

The standard super sold by bee supply manufacturers today is designed to hold ten self-spacing frames. In practice, most beekeepers use only nine frames in a ten frame super. When the foundation is first being drawn it is advisable to have ten frames in the supers; otherwise burr and brace comb will be built between the sheets of foundation.

The width of a top bar of manufactured frames is 1 1/16 to 1 1/8 inches. The use of wide top bars dates from the comb honey era. When nine or ten such frames are crowded together in a brood nest the bee space between them is minimal, about one-quarter of an inch. Excluders are not used in comb honey production. Frames close together in the brood nest below the

EQUIPMENT FOR COMB HONEY PRODUCTION 19

comb honey supers tend to act much like an excluder and deter the queen from going into comb honey sections. Under extremely crowded conditions queens have been known to lay in comb honey sections and of course this is not desirable.

In a normal colony, pollen is stored above and along the sides of the brood nest adjacent to the brood and between the brood and the honey. Pollen has a bitter taste, in addition to being unsightly in a comb honey section, and its presence in sections is to be avoided. Wide top bars appear to encourage bees to store pollen in the brood nest area and not in the sections, another reason for using them.

I'm not certain that modern-day bee supply manufacturers understand why they make their top bars so wide. In fact, insofar as extracted honey production is concerned it would be preferable to have frames with top bars 7/8 to 1 inch wide. I am aware that some commercial beekeepers who make their own frames have top bars only 3/4 of an inch wide; this allows for better ventilation.

I prefer to use self-spacing frames when producing comb honey. Beekeepers have used staples and nails for spacers on home-made frames which have the end bars the same width as the bottom bar and top bar. The chief disadvantage in using nails and staples is the fact that one might hit them with an uncapping knife when producing extracted honey; other than this they work quite well.

Most beekeepers use pine for making frames. The chief reason is that pine wood is less inclined to split when the frames are nailed and assembled. Pine is easily nailed. Hard wood is satisfactory for frames but nailing it is difficult. Where the wood is unusually soft, metal eyelets should be used when wires are placed in the frames. All of the wood which is used on the interior of the beehive should be sanded or well planned. Bees prefer to have smooth surfaces in the hive, and will varnish rough areas and fill cracks with propolis.

Follower boards

Follower boards, sometimes called dummies or dummy boards, are flat boards the size of a frame but only 1/4 to 3/8 of an inch thick. They are little mentioned in the literature today and are not sold by any manufacturer. However, they were very popular many decades ago, especially with comb honey producers. Some beekeepers prefer a follower board which is slightly shorter than a standard frame. They were popular with beekeepers who use both eight frame and ten frame hives. Most beekeepers use a single follower board on one side of the hive, though occasionally some used two on both sides. The standard dimensions would be 17 5/8 by 9 1/8 inches, not including the lugs.

The chief advantage of the follower board is that it is more easily removed than a frame. Comb honey producers check their brood nest for queen cells frequently. It is important that they be able to move rapidly and that bees not be killed in the process of moving frames and inspecting the colony.

In our own operations we use follower boards only rarely, but I recommend that beekeepers give consideration to them.

Covers and innercovers

One could write a long chapter on covers and innercovers. There is no perfect way of covering a hive. As with many other pieces of equipment, which beekeepers use, if there were a perfect hive cover everyone would use it.

Commercial beekeepers interested in liquid honey production often use combination cover-bottomboards. These have the advantage of being interchangeable. They are usually light in weight and save both weight and space for those who migrate with their bees. Combination cover-bottomboards are easily made and cost much less than a standard cover and innercover. However, insofar as comb honey production is con-

EQUIPMENT FOR COMB HONEY PRODUCTION 21

cerned, I prefer a standard telescope cover and innercover. The hole in the innercover should be covered when it is being used above comb honey supers (see Chapter VII, Closing upper entrances). During the remainder of the year, we prefer to leave the hole in the innercover open, as bees will deter ants and other insects from nesting in the space between the cover and innercover.

In a properly made super there is one quarter inch above the frame and the level top of the super and one eighth of an inch below the frame and the level bottom of the super. In the case of a comb honey super the bee space is above the sections only. This means that when an innercover is placed above the top comb honey super, there is only one quarter of an inch between the tops of the sections and the innercover. This is minimum bee space. Therefore it is important that the innercover not be warped and that the top of the sections, when placed in the section super, are perfectly level.

During the time comb honey is being produced the colony should be perfectly level. This matter is discussed in greater detail below. Water often does not drain well from level combination cover-bottomboards, another reason for using the old-fashioned telescope-type cover.

Hivestands and grass

It is poor beekeeping practice to have colonies directly on the ground. Bottomboards in contact with the ground are often wet and wet bottomboards tend to cool the hive and make the brood rearing temperature control difficult. Colonies which are close to the ground may have entrances which are obstructed by grass or other debris.

Beekeepers have used a great variety of permanent and temporary hivestands, including old tires, tin cans, bricks, cinder blocks, slab wood, cement, iron, etc. I prefer to use a hive stand made of 2 by 6 inch lumber. The cross members

2. Active colonies after unpacking in the early spring. The entrances are still reduced with cleats. The hive stand keeps the colony bottom boards off the ground and dry. The fact that the supers are painted different colors tends to deter drifting.

of the hivestand are spaced so that when two colonies are immediately adjacent to each other for winter packing there is a dead air space below the hivestand. Hivestands treated with creosote or a 24 hour cold soak with pentachlorophenol will have a life of 20 to 25 years.

In recent years I have found it expedient to cut pieces of roofing paper, approximately 3 by 5 feet, to place immediately under the hivestand. This prevents grass from growing around the hives where it can be a nuisance; it also makes mowing the bee yard much easier. One may use worn winter packing paper in much the same way.

3. Cinder blocks make good hive stands and will also slow the growth of grass around a colony entrance as is shown here. The block, in part, serves as an alighting board though honey bees do not need a large alighting area in front of their hive.

4. Bees will often survive under adverse conditions, but in our experience a colony which rests on the ground, as this colony does, will have a damp bottom board which tends to cool the hive.

The great argument — The best size hive

In my opinion management is more important than the size of hive or race of bees which one uses in honey production. Probably 90 per cent of the beekeeping equipment in the United States is of the 10-frame Langstroth size. Standardization is worthwhile for two reasons: It protects one's investment and makes resale easier. If one buys additional equipment it is interchangeable with that which is on hand.

L. L. Langstroth made his hive the size it is because of the dimensions of the lumber which existed at the time. While Langstroth was concerned about the size of the hive in which to keep bees, it was not a major consideration. Generally speaking, between 1880 and the First World War, there was a strong preference in the United States for the 8-frame hive.

During the First World War there was a great increase in the number of hobby and beginner beekeepers. At this time Drs. E. F. Phillips and E. R. Root concluded, and probably rightly so, that beginner beekeepers would have better results wintering colonies in two standard, 10-frame Langstroth hives than they would attempting to overwinter a single 8-frame hive. At the time many commercial beekeepers were successfully overwintering their colonies in a single 8-frame brood nest, feeding the colonies sugar syrup late in the fall. The method is an excellent one but it requires attention to detail. Also, many beekeepers were successfully wintering their bees in cellars at the time. Colonies were usually overwintered with only 20 to 30 pounds of honey or high quality sugar syrup; this is an obvious saving over the present system. In their enthusiasm Phillips and Root also convinced many commercial beekeepers to change their wintering systems, and today it is most popular to winter bees in two 10-frame Langstroth supers with 60 to 80 pounds of honey.

In this whole question we are concerned with two markedly different problems: First is that of overwintering and second

EQUIPMENT FOR COMB HONEY PRODUCTION 25

is that of honey production. Insofar as honey production is concerned, we want colonies to grow rapidly and to their full potential in the spring. Insofar as wintering is concerned, we want the colony to survive with a minimum consumption of honey.

Carl E. Killion says (on p. 13 of his book, *Honey In The Comb*) that he prefers the 10-frame hive. His studies of 8- and 10-frame hives show that over a three-year period he made more finished comb honey sections on his 10-frame hives than he did on his 8-frame hives. This is curious because in the same paragraph he states that the two comb honey producers he admired the most, Dr. C. C. Miller and Mr. Charles A. Kruse, both used and preferred 8-frame hives. Interestingly too (on page 27 of the above book), Killion pictures an 8-frame comb honey super which he places on top of one of his 10-frame hives. In the picture caption he states, "the narrow super permits the sections to be directly above the brood at all times". Killion is not the only one who prefers to use narrower supers on 10-frame hives. One of the New York State's better and more famous honey producers, Clarence Schraeder of Waterville, N.Y. uses the narrower 8-frame size comb honey supers above his 10-frame hives.

I have discussed this question at length with Mr. Gerald Stevens, the State Apiary Inspector for New York State. Both his father and grandfather were commercial beekeepers in Cayuga County where they operated 1200 colonies for many decades. The Stevens made a switch from 8-frame to 10-frame hives at the end of World War I. Mr. Gerald Stevens said that many years later his father made the remark that "that was the end of 80-ton crops." This meant that when the family was using 8-frame hives they had made more honey on the same number of colonies than after they made the switch. Such observations are, of course, subjective. It is entirely possible that as much honey was produced but that more honey was left with the colonies for winter. At the same time, we are well aware of the fact that agricultural patterns in Cayuga County were changing drastically; this alone could account for the change.

I'm also strongly influenced by some observations being made by Mr. Thomas Seeley, a graduate student at Harvard University. His data, which should be published about the same time as this text, show that bees live successfully in bee trees in cavities which are, on the average, smaller in size than the space in a standard 10-frame Langstroth super. Seeley is the first to study the natural nest of the honey bee so extensively. In matters of this sort it is well to be informed about what honey bees really do and not what we think they do. I suspect there will be more research on this subject in the near future.

Summary: Insofar as honey production is concerned, the 10-frame hive, at least for the brood nest, is probably best; insofar as wintering is concerned, more research is needed. Wintering a single 8-frame hive may have advantages.

II.
Supers and Sections

Manufacturers of bee supplies sell only three sizes of comb honey sections today. These are the 4 1/4 by 4 1/4 by 1 7/8 inch beeway sections which are most popular; the 4 by 5 by 1 3/8 inch sections which are used by only about 15 percent of comb honey producers; and the 4 1/4 by 4 1/4 by 1 1/2 inch plain sections which are least popular. In years past some beekeepers have made sections which would hold two pounds of honey. The smallest sections I have ever seen held three to four ounces of honey. The outside dimensions were approximately 2 1/8 by 2 1/8 inches; these have not been available since the early 1950's. They are often referred to as the quarter pound sections. When properly filled they make a delightful gourmet individual package, suitable for the finest restaurants. I regret they are no longer available.

My personal preference is for the taller 4 by 5 sections. They require about one ounce less honey to be properly filled, but more important, a greater percentage of 4 by 5 sections will be properly filled and finished. In comb honey production the sections must be centered above the brood nest and those which are not will not be properly filled. Sections in the outside section holders and the cells in sections near the outside of the super are the last to be drawn and filled with honey. More important, perhaps, it is obvious there is more empty space in a 4 by 5 section super, and ventilation by bees is therefore easier. Because of the decreased interest in comb honey production in recent years, manufacturers have told me that it is entirely possible that the 4 by 5 section will not be available in future years. Its demise would be a great loss.

As is the case with sections, manufacturers offer only two sizes and types of comb honey supers. While very few beekeepers make their own sections, it is practical to make one's own comb honey supers and furniture. In the beginning of the comb honey era sections were made from four pieces of wood which were nailed together. Dovetailed sections are a more recent innovation; four piece sections are not longer available.

5. A standard, factory-made 4¼ by 4¼ section super filled with split sections as it appears when it is removed from a hive. Boards one half inch thick have been placed along the sides; this centers the sections over the brood nest.

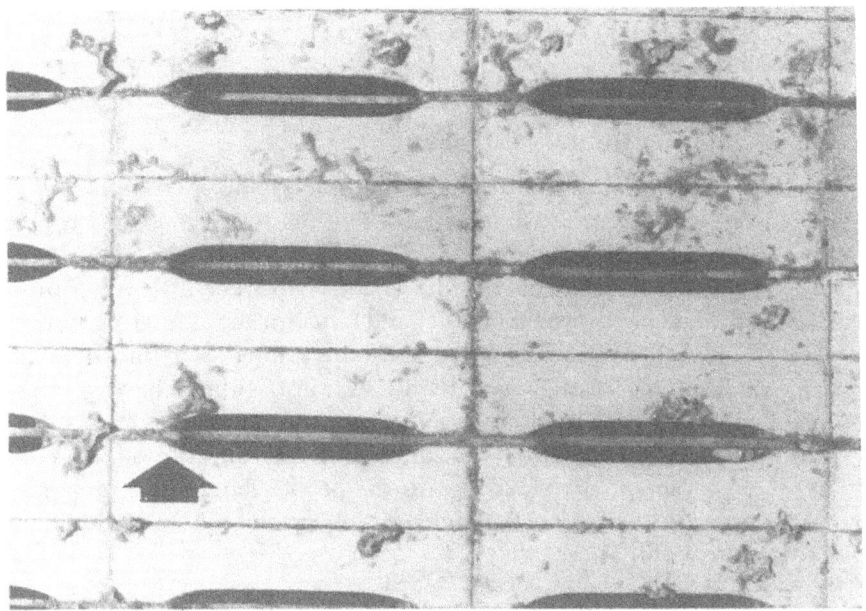

6. *This close-up of square, factory-made, split sections shows the points (arrow) at which bees deposit propolis in excess. The bees have built burr comb on the tops of most of these sections; this is easily removed with a sharp knife; however, less burr comb would have been built if the tops of these sections had been coated with paraffin.*

Split versus solid sections

Split comb honey sections, those which are split on three sides, are a recent innovation. People who use solid comb honey sections must fix a piece of foundation into each section individually. This is usually done with molten beeswax. The split section has the advantage that a piece of foundation may be put into four sections simultaneously. A special device for holding the split sections apart is required but with a little experience the job is easily done. Only certain types of supers will accommodate the split comb honey sections; these are discussed below.

30 COMB HONEY PRODUCTION

I think the purists among the comb honey producers prefer the solid sections. Some have said the slit in the wood is not aesthetically pleasing; I don't really think it makes that much difference. The use of split sections requires a little less equipment and is probably easier and faster for the average producer. It is sometimes difficult to fit the foundation into the end sections, in the case of split sections, without having some warping of the foundation; however, with practice it can be done. Manufacturers of foundation are not always as careful as they might be as regards the proper length of foundation for split sections; that which is too long or too short should be rejected.

Supers

One who has visited beekeeping museums, or even a beekeeper who has been in business a great number of years, is aware of the great variety of beekeeping equipment which has been built. Some of this was made with a notion as to what someone thought bees would prefer, but much of the equipment was a result of long years of testing by successful beekeepers. In surveying what comb honey producers have preferred, one soon realizes that there has been a strong preference for the "T" super and the 4 by 5 inch comb honey section.

Comb honey super furniture is made of pine. As in the case of frames this is simply because soft pine nails easily, can be sanded to a smooth finish, and cleans without too much difficulty. Cleaning comb honey furniture is a chore which every beekeeper must face and those who produce large quantities of comb honey have often chosen the type of furniture which is most easily cleaned. Clearly, the slatted type fences or separators which are used with the taller 4 by 5 sections are not so easily cleaned as are the plain separators used in the Killion super.

7. Several things are unique about this homemade comb honey super containing 4 by 5 inch sections. This super, which is a modified eight-frame super, has wide edges so that it may be placed on a ten-frame hive. Note the crawl space on the outside edges, away from the sections. The homemade fences between the sections do not come all the way to the top of the sections, yet provide a perfect bee space and allow for a maximum of ventilation. This super was made and used by Clarence Schraeder, who until he retired, operated 400 colonies for comb honey production in central New York State.

The "T" super, especially as modified by Killion, has the added advantage of good ventilation and a space for bees to move on the end of the super. Killion states that this cuts down on the deposition of propolis, and I believe he is correct. The small "T" tins which are used in the super are usually cleaned by dunking them in a boiling lye solution.[1]

---

[1] Beekeepers have used boiling solutions of lye water rather successfully for a great number of years. However, lye is caustic and a dangerous material to use. The lye should be placed in cold water and the water then brought to a boil. If lye is poured into boiling water, a near-explosion will take place; if one spills lye on his body or clothing it should be removed immediately with cold water or serious damage may result. (For more on using lye, see chapter VI, Cleaning the frames.)

32 COMB HONEY PRODUCTION

The Cobana super and sections

One of the very few worthwhile recent discoveries as regards beekeeping equipment is the Cobana comb honey section super, which was invented some time during the 1950's. Cobana comb honey sections are round; they are four inches in diameter. They hold eight to nine ounces of honey. Most of the Cobana sections I have seen in honey shows are completely capped and have a very fine physical appearance. The corners which exist in normal comb honey sections are missing and, of course, it's the corners which bees have difficulty filling.

The furniture for a Cobana comb honey super is plastic. The inventor was fastidious in constructing the correct bee space, and because he was careful in this regard one very rarely sees any burr or brace comb on the furniture. The plastic rings which contain the final product are completely protected by the furniture, and thus the beekeeper is spared the time and trouble of cleaning the sections as he must do in the case of wood.

So far as I can determine, the Cobana system has never been patented. However, the fact that the molds necessary to make the plastic furniture and Cobana rings are expensive is probably ample protection for the inventor. The sales of the furniture and material to make Cobana sections have been limited, and these items are not available through the normal distributors.

I suggest that if someone were to market a modified version of the Cobana section, or if the original sections and furniture were widely advertised, that Cobana sections would soon replace wooden sections. The finished sections can be retailed for essentially the same price as a standard comb honey section, and since they contain less honey and the hard-to-fill corners in a normal section are eliminated, the beekeeper has less difficulty in obtaining quality Cobana sections in quantity.

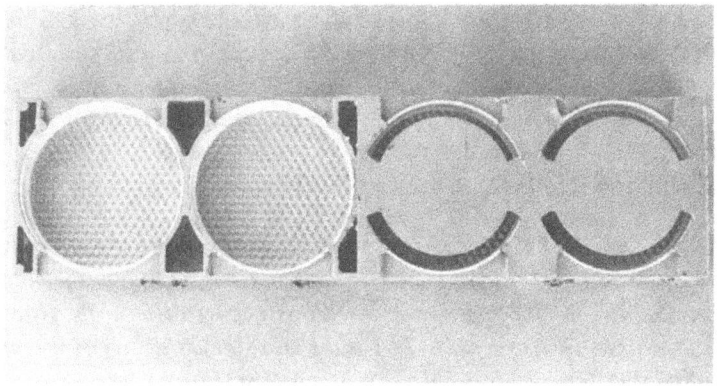

8-9. *Cobana rings and section holders. The section holders are made in such a way that the fence, which is a permanent part of the section holder, is not continuous for all four sections on one side of the holder. It will be noted that the bees have a minimum contact with the section rings, making preparation for market simple.*

34 COMB HONEY PRODUCTION

The cell walls built inside the Cobana rings are visible through the clear plastic and are probably not too attractive since they are never uniformly straight. However, when marketed, Cobana sections have a wrap-around label which covers the cell walls and holds the covers for the sections in place.

Judging Cobana sections at honey shows has always been

10. A Cobana section with broken cappings. This section was on the side of the super and the wrong fence was used. Side fences in the Cobana supers are constructed differently from those used in the center. The same problem may occur when improper fences are used with wooden sections.

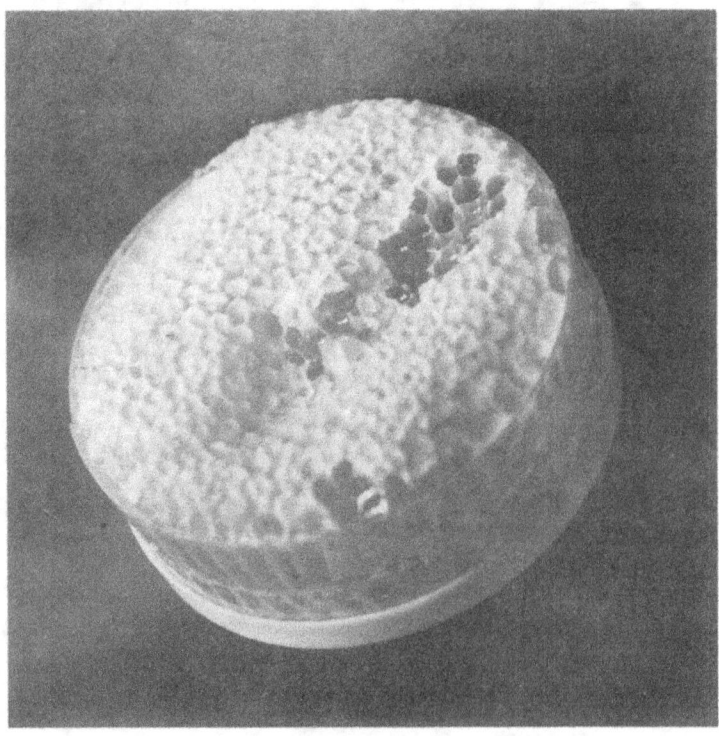

difficult. I have urged superintendents to create special classes for Cobana sections because I do not feel it is fair to judge them in competition with regular sections. This is because the design of the Cobana section is nearly perfect, in my opinion, and there are few errors a beekeeper can make in their preparation for market. Thus, most Cobana sections score rather high in shows.

In all regards the management involved in the use of Cobana sections is similar to that which is used to produce normal comb honey sections. While it is true that the colonies probably do not need to be so crowded as when producing normal sections, the crowding of colonies will result in a greater production of Cobana sections. It is regrettable that they are not more widely known and used in beekeeping circles.

Foundation

Foundation is made by embossing the 6-sided, 3-faced cell base onto thin sheets of beeswax. Bees normally, but not always, build straight, level combs. The use of foundation enables the beekeeper to better control the resulting frames and combs.

A glance at a bee supply catalog shows that foundation is available in various weights. Medium brood foundation, used for standard frames, will have more sheets of foundation per pound than will heavy brood foundation. Only thin super foundation should be used to produce comb honey and cut comb honey. The same rolling mills, which look not unlike an old washing machine wringer, are used to make all types of foundation. However, that foundation which is used for comb honey production is made from pieces of beeswax which have been rolled so as to make the wax foundation thinner.

Since people who eat comb honey also eat the foundation or midrib, it is important that it not be so tough that it is objec-

36 COMB HONEY PRODUCTION

tionable to them. One has only to take a mouthful of comb honey that was cut from a frame made with a heavy or medium brood foundation to appreciate the difference in chewing quality.

Starters versus full sheets of foundation

Most beekeepers, whether they are making combs for the brood nest, extracting frames, or comb honey sections, prefer to use full sheets of foundation. The reason for this is to obtain a uniform comb with worker size cells. However, as many beekeepers are aware, it is possible to use strips of foundation known as "starters" and to obtain perfect combs. In the case of comb honey sections it is generally agreed that sections made with worker size cells are more attractive than those made with drone size cells.

Whether bees build comb with worker size cells or drone size cells depends upon several circumstances. Large, populous colonies are more likely to build drone comb than are smaller, weaker colonies. However, comb building is also influenced by the honey flow and there appears to be a tendency during the good honey flow for bees to prefer to build the worker size cells. I prefer to use full sheets of foundation both for sections and frames, but suggest this is an area where individuals may experiment.

Odd size cell diameters

In certain parts of the country, notably Arizona and Southern California where it is unusually dry, the honey which is produced will have a lower moisture content than it does further north. It is interesting that honey containing 16 per cent moisture flows much less easily than that which contains 17 or 18 per cent moisture. Beekeepers in arid areas soon found that it was much easier to remove honey from combs with larger cells with an extractor than from combs containing smaller cells which would be worker size diameter. Foundation with intermediate size cells has also been designed, again, largely from the point of view of facilitating extraction. However, from time to time, there has been discussion of using foundation with larger cells or intermediate size cells for comb honey production. As indicated above it is generally thought that sections of comb honey made with worker size cells have the greater eye appeal. Other than in dry areas in the United States I advise the use of foundation with standard size cells only.

When to put foundation into sections

Beeswax is very brittle at temperatures near freezing and melts at approximately 148°F. Similarly, it expands and contracts with changes in temperature. Thus, if one installs sheets

11. A brood comb with excessively shallow and deep cells caused by warped foundation. The drone cells, in the top center of the comb, cover foundation which warped away from the center of the comb, and the viewer; the shallow cells in the bottom center were caused by foundation which warped toward the viewer. Under ideal conditions foundation should not be put into frames or sections until they are to be placed on colonies.

38 COMB HONEY PRODUCTION

of foundation in frames or comb honey sections when the temperature is 70°F. and then stores them in a room where the temperature occasionally reaches 80-85°F., or the reverse, the foundation will warp and buckle. Many of the poor combs which exist result from beekeepers placing warped and buckled foundation on colonies prior to the honey flow.

From an ideal standpoint, foundation should be placed in frames or comb honey sections and put onto colonies immediately. If the honey flow is in progress the bees will start to draw the comb and perfect sections will be made. From a practical point of view it is not always possible to put the foun-

12. Many things are wrong with this unmarketable section. The foundation was not properly fitted into the end of the split section on the left; this warped the comb and the bees built cells against the fence. The open cells in the comb indicate further warping. The section is propolized excessively and while it may be possible to clean the wood, it is obvious the section was left on the colony too long.

dation into sections at the last minute. Where sections or frames of foundation must be stored it is important that this be done in a room where there is good temperature control. Beekeepers who do not exercise control in this regard will have poor combs and sections. Many commercial comb honey producers were in the habit of employing extra help on those days when colonies were being supered for the express purpose of there being no delay between the time the sections were prepared and when they were put on the colonies.

Mounting blocks and methods of fastening foundation

Foundation is placed in solid sections using a board covered with blocks which fit inside the folded sections and which are half their thickness. The foundation is cut into small pieces prior to fixing it in place. The foundation is first fastened to the top inside of the section. There are basically two ways of fastening the foundation in place: The preferred method is to use a hot iron which melts one edge of the foundation which

13. *Sections which are not split are fitted with foundation by placing the section over a block just slightly smaller than the dimensions of the section. The foundation may be held in place by molten beeswax or paraffin. Some beekeepers prefer to fit the section boxes with two pieces of foundation as is shown on the right; however, this extra precaution is probably not necessary.*

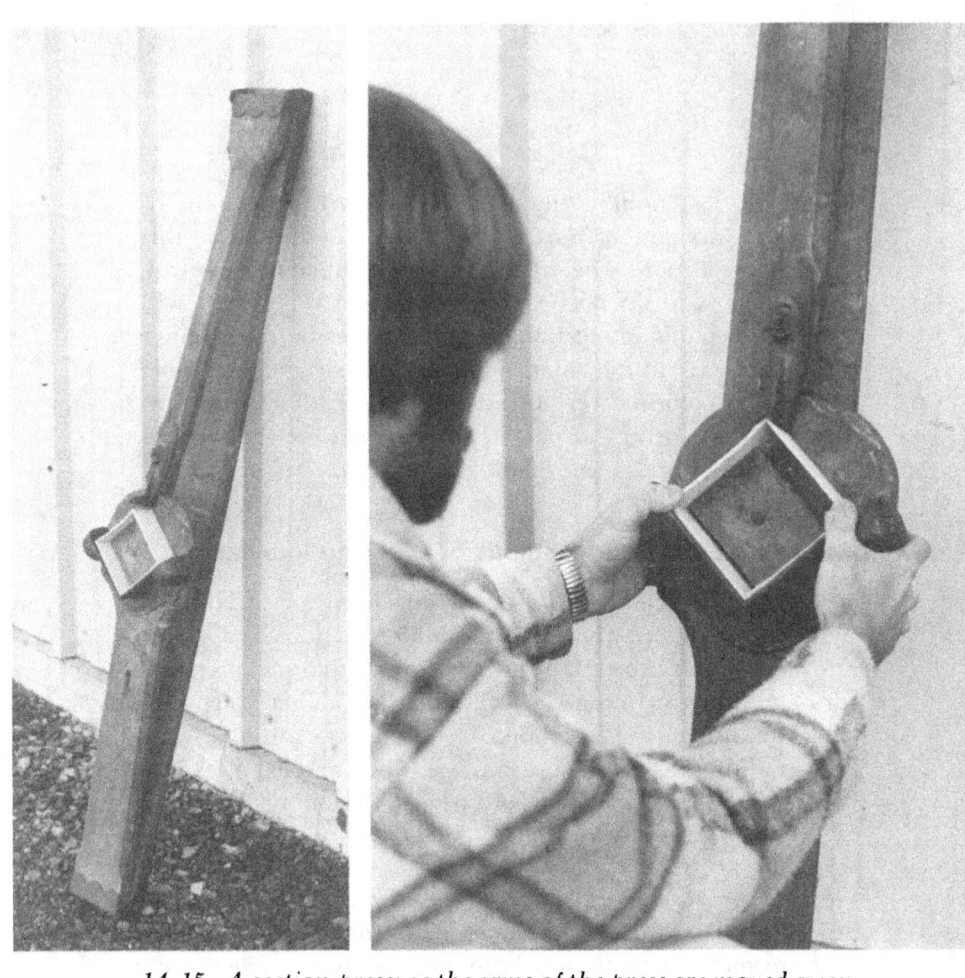

14-15. A section press; as the arms of the press are moved away from the operator the dovetails at the top of the section are pressed firmly into place. Using a section press has the advantage of making square sections.

is then pushed against the edge of the section as the iron is withdrawn. One must be careful not to melt too much wax. When properly done one whole edge of the foundation adheres to the top inside of the section. An alternate method is to use a small brush and hot molten beeswax or paraffin. One needs only a few drops to hold the foundation in place. In addition to

SUPERS AND SECTIONS   41

16-17. The easiest way to place foundation in split sections is to use the metal device pictured which holds the tops of the sections apart about one quarter inch while the foundation is put into place. Care must be exercised to insure that the foundation is fitted properly into the ends of the outside sections.

18-19. *A homemade device for spreading split sections so that they may be fitted with foundation. These sections were wetted with water about five minutes before they were folded; that is apparent here. It will be noted that the sides are warped slightly near the top. This will be corrected when the sections are placed in the section holder, but there may be a small problem with warped foundation. These sections show that it is preferred to keep the sections in a damp atmosphere or room so that they will not warp when folded.*

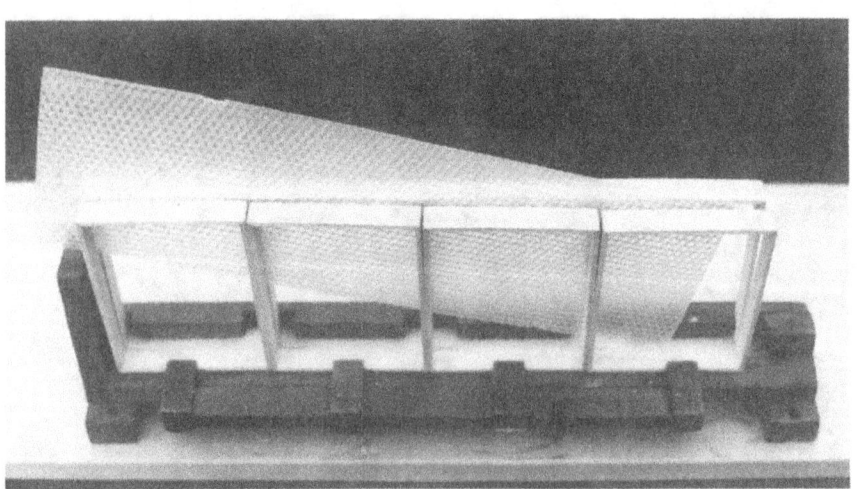

fixing the top of the foundation to the inside of the section, one side of the foundation should also be fastened in much the same manner.

It has been popular with some comb honey producers to use two pieces of foundation in a section. Carl Killion advocates and illustrates this method in his book on comb honey production. When two pieces of foundation are used, the larger of the two is affixed to the top and one side of the section. The second piece of foundation which is about half an inch wide and four inches long is affixed to the bottom of the section. The reason for using two pieces of foundation is that it is thought the bees will do a better job of filling the corners of the section, especially those along the bottom. However, in my opinion this is also a function of the honey flow and in a good honey flow the bees will fill the whole section about equally regardless of the foundation.

Placing foundation in split sections is faster but in some ways more difficult. This is especially true as regards fixing the foundation to the outer edges of the two outer sections in each section holder. If the foundation is not held firmly in place in a split section, it may fold or buckle and the result may be an unmarketable section. Metal devices for holding and spreading split sections are available and should be used. If the foundation does not fit into the outer section, in the case of split sections, it may be possible to use a small amount of hot wax or paraffin to hold them in place to prevent their warping.

Preparation of sections and supers

Only a few companies make sections for comb honey supers today. Those which are made are of high quality and are manufactured from basswood. Basswood lumber has long, white, flexible fibers which gives a fine finish when sanded.

Basswood, unlike many woods, has little odor and thus it is not objectionable to bees, nor does it impart any odor to the comb honey which might be offensive to man.

Even perfectly made comb honey sections will break at the corners if the wood is too dry when the sections are folded. For best results the sections should be stored in a room with high humidity for several weeks before being prepared for use. Placing a wet blanket or several wet towels over the box containing the sections will help in this regard. Alternately, one may remove the sections and spread them on a long table, wetting both the upper and lower surfaces, along the folds. The sections will be ready to fold about five minutes after they are wetted. Wetting the wood will cause some of the fibers to rise and gives the exterior surface a slightly rough texture; for this reason storing the sections in a high humidity room is to be preferred.

Various gadgets for speeding up the folding of sections have been devised. A properly made section press has the advantage of giving its maker a square section though in large part the way in which the section is held in the super is responsible for its final shape.

Comb honey super furniture is designed to allow the bees to have a minimum of contact with the wood surrounding the section. Since the wood is part of the final package it is desirable to keep it as free of propolis and travel stain as is possible. When the super is finally assembled all the parts should be squeezed together as tightly as possible to protect the wood. Super springs, made of strong steel, are used in the final assembly and make the preparation of the supers easier.

When the super is finally assembled, the tops of the sections are exposed to the bees; this cannot be avoided. Old-fashioned comb honey producers added (painted) a light coat of paraffin to the top of their sections at this stage and this was later scraped from the sections with a knife before they were marketed. Few beekeepers paraffin the tops of their sections today, but I think it is a desirable practice. One may use beeswax though it more expensive. The paraffin or beeswax is usually put on these sections with a one-inch-wide lightweight paint brush. Only a thin coat of paraffin needs to be used. Those people who do not paraffin their sections attempt to keep the supers on the hives for a minimum period of time, thereby reducing travel stain on the top of the sections.

Queen excluders

Comb honey producers need not use queen excluders. Queens have been known to move into comb honey section supers and to lay eggs in individual sections, but they do so very rarely. If, as above, frames with wide, thick topbars are used and are properly spaced in the brood nest, they will tend to keep the queen in the super below the comb honey supers. Also, the location of comb honey supers on the colony, as discussed below, will tend to discourage the queen's presence in sections.

Insofar as extracted honey production is concerned, my personal preference is to use queen excluders but to place them on the colonies about three and a half weeks before the honey is to be removed. Queen excluders inhibit the movement of bees only slightly. (If one observes workers moving through excluders closely it will be noted that they do so with considerable ease.) The chief disadvantage of a queen excluder is that it makes ventilation of the colony by the worker bees more difficult. Under crowded conditions it therefore encourages swarming.

III.
The Apiary Site

In every state there are primary, secondary and marginal beekeeping areas. During the past 30 years the trade journal, *Gleanings in Bee Culture,* has carried a series of articles which discuss the beekeeping potential in many of the states. Maps accompanied most of the articles which were authored by the extension apiculturist, state apiary inspector or a knowledgeable beekeeper in each state. Most state college libraries have back copies of the trade journals. These can be searched for pertinent information about the honey producing potential in an area. Successful commercial beekeepers move to those areas where honey plants abound. It is not practical for beekeepers to plant plants for honey production. Since there is usually a premium for the lighter, milder honeys, beekeepers tend to go to those areas where plants which produce this kind of nectar are present. Most honey consumers prefer honey made from alfalfa and the clovers. There are, of course, exceptions, and in some parts of the country a good market exists for such strong flavored honey as buckwheat, goldenrod, wild thyme, tulip poplar and even honeydew.

A single colony of bees, with no competition in the vicinity, may survive almost anywhere in the United States. Some beekeepers have pursued their hobby by feeding colonies during the off season in some of the mountainous and arid regions. However, from a very practical point of view, it is important that beekeepers ascertain how many colonies their territory will support and then not overstock it.

In a few towns and villages along the east coast, a small number of beekeepers have found it difficult to produce a high quality honey because of the presence of privet, which has a nectar and resulting honey with an objectionable flavor. Privet is a popular hedge and when the plant flowers bees are attracted to it in great numbers. When one has bees in such an area the only recourse is to move the bees. In contrast to this situation, some towns and villages have lined their streets with basswood or linden trees (Malden, Massachusetts is one such example). Because of the great number of trees planted in some locations, beekeepers have found that they can secure a crop of very delicious honey in an area which might otherwise not be good beekeeping territory. Of course, there are limits to the number of colonies which can be kept in an apiary under any circumstance.

Commercial Sites

Men who make a living keeping bees feel that they can have a profitable operation only when they have forty or more colonies per apiary. This is not possible everywhere. Such an apiary will usually occupy one-quarter to one-half acre, though if the site is surrounded with bushes and trees, as it should be, a total of one to two acres may be required.

Commercial beekeepers insist that their apiaries be accessible by road. It is time consuming to carry supers and equipment even a few hundred feet. Commercial beekeepers will often spend many weeks each year repairing their access roads, filling ruts with stone and gravel and cutting trees and brush which are in the way.

Keeping bees in congested areas

Beekeepers have an obligation to make certain their bees are not a nuisance. Bees have been kept successfully in congested areas and long articles have been written about the precautions which should be taken so that a beekeeper's neighbors are not disturbed. However, the considerations may be reduced to a few sentences: The apiary site should be surrounded by a hedge or fence at least six feet high so the bees are forced to fly above the heads of people in the vicinity. Bees should be provided a source of water within the apiary or close to it so they will not collect water from bird baths and swimming pools in the vicinity. Bees vary in their temperament and beekeepers in congested areas should keep only gentle bees; when colonies are found which have more aggressive bees, they should be requeened immediately. Beekeepers in congested areas should work their colonies only on warm, sunny days when the majority of the bees are in the field gathering nectar. It is the older bees which are more agressive and if they are away from the colony at the time it is being manipulated there is much less chance the beekeeper and his neighbors will be stung. If these simple rules are followed a beekeeper will have no difficulty pursuing his avocation.

In towns and villages where ordinaces do not allow honey bees to be kept, it is my opinion that the number of stinging insects does not decrease after the bees are removed. If one removes a colony of honey bees from an area, he makes the pollen and nectar they would consume available for other insects and as a result their numbers will increase. Many insects feed on nectar and pollen but the solitary bees and the social wasps are especially common around flowers in areas where there are no honey bees. It has been our observation too that over half of the honey bees found in urban areas live in the sides of houses, garages and barns. Not even new houses are exempt from the problem. Bees need only to find a small entrance to a cavity to make a new home.

The best way to eliminate stinging insects from a town or village where they are not wanted would be to forbid the growing of flowering plants. Most of the insects which sting are dependent upon flowers for their food and cannot survive without them. In the case of the social, garbage-feeding wasps, even this would not be sufficient, for certain of these species thrive around fast-food establishments and parks where they feed on refuse; a strict sanitation code must be enforced to rid these areas of stinging insects.

At the present time only seven per cent of our population lives on farms. The food which all of us consume is produced by fewer than two per cent of our population. A beekeeper has an opportunity, and perhaps an obligation, to point out to his friends and neighbors the important role bees play in our economy. Frequent discussions, and the occasional gift of a jar of honey will do much to calm the fears and suspicions of neighbors who do not understand bees and beekeeping. Still, in the final analysis, it is the beekeeper's obligation to make certain his bees are not a nuisance.

Physical requirements of good apiary sites

A good apiary site has a maximum of sunlight, is level or slopes to south or east, has good air and water drainage, is accessible and has a supply of fresh water nearby. A windbreak is helpful; a woods, hedge or fence not only protects one's neighbors and hides the bees, but reduces the wind in the vicinity of the apiary. These requirements are important both during the honey flow and remainder of the year, especially in winter. I prefer to have a few small trees or bushes within a beeyard to serve as landmarks but this is not necessary.

One often sees apiaries in the woods or in heavily shaded

areas. Colonies which are shaded require a greater percentage of their work force to warm the hive and to keep it dry; as a result there are always more bees in the hives and even on a warm day such colonies will be more difficult to work because of the greater number of stings received. In the Arizona desert, where the temperature may rise to about 120°F., I have seen colonies shaded; I presume this is done in a few other areas too. In Florida, and the southern states, most beekeepers paint their hives white or with aluminum paint to reflect the heat; however, in the northeast I know of one beekeeper who paints his hives black so they will absorb the heat. Our University colonies are painted a variety of colors including green, yellow, blue etc. Painting hives different colors helps the bees to identify their own hive better and to reduce drifting.

Most commercial beekeepers mow their apiaries once or twice a year. This is done mostly for their own convenience. An advantage of keeping bees in some of the arid areas, such as parts of California, is that this is not necessary. Where sites need mowing it is well to remove rocks and debris. I have kept colonies in pastures. Sheep do a good job of clearing the grass around beehives; however, horses and cows will tip over the occasional hive, especially when they are stung. They do not tip hives on purpose, or with a vengence, but merely because they become excited and kick and run aimlessly when stung. Apiaries where larger animals roam should be fenced to keep them out.

A minor point which is discussed in greater detail below concerns having a level site for the colony itself. Under normal circumstances it is recommended that colonies tilt forward slightly. This is done so that water will drain from the bottomboard and there will be less accumulation of ice and snow at the entrance. Producers of liquid honey leave their colonies in this position all year. However, those who produce comb honey must level their colonies while the bees are in the process of filling sections (see chapter VII, Leveling colonies).

Outbuildings

Producing comb honey is fussy work. In the case of liquid honey production one need not be so concerned about having every frame full and using perfect, hole-free equipment. Honey flows are often short and repairs when producing comb honey cannot be delayed.

One can give many reasons why having a building in an apiary, when the site is not in one's backyard, is an advantage. It is helpful to keep a few tools, nails, pieces of wood, cork, smoker fuel, paper, extra supers, bottomboards, etc., on hand at all times. Some beekeepers will rearrange sections in a comb honey super, especially near the end of a flow. It may often be helpful to move partially filled sections from the outside of a super to the middle where they will receive more attention from the bees. Perhaps a book on comb honey production is not precisely the place for a discussion of outbuildings; however, I have found them useful and recommend them to all beekeepers — comb honey producers will find an outbuilding in their apiary especially helpful. Because all this is true, many beekeepers produce comb honey in their backyards only where their work bench is nearby.

IV.
Outline of Seasonal Management for Comb Honey Production

Management is the key to maximum production. Management includes having good equipment which is cleaned routinely (scraped free of propolis, burr and brace comb), good queens, good combs, etc. However, in this chapter we are specifically concerned with those things beekeepers do on a month by month, week by week basis.

This chapter is based on my experience with comb honey production in central New Yorf State. Beekeepers farther north and south will need to adjust their schedules accordingly. Spring management begins with the flowering of pussy willow which takes place about the middle of April; apples bloom in New York State between May 20 and 30. Clovers start to flower about June 15 or 20, but for us in the Southern Tier of New York State this is not a major honey flow. Basswood blooms July 7 or 8; goldenrod, and there are several species of it, is first visible the last few days of July, and our goldenrod honey flow may start any time between August 1 and September 20. These plants are mentioned in the discussion below; beekeepers not at the same elevation or latitude may tie much of their management scheme to flowering of these plants in their area.

A thorough knowledge of one's local honey plants is neces-

MANAGEMENT FOR COMB HONEY PRODUCTION 53

sary to be a success in the beekeeping business. Of course, just because these plants exist in an area does not mean they will always produce a surplus of pollen or nectar; the quantity of nectar produced is a function of the soil and the weather.

March-April — unpacking and the first inspection

Beekeepers seldom examine their colonies in winter; doing so does less harm than one might think, but it is really not necessary. Apiaries should be visited occasionally during the winter to check for vandalism and damage by rodents and mammals. The real work starts about the time the pussywillow and skunk cabbage flower. In our area, especially in cities and villages, people have planted so many crocuses that they are becoming increasingly important as a source of early spring pollen. Crocuses bloom a few days ahead of pussywillow and may serve to stimulate colonies earlier in the year. Whenever the first flowers bloom and bees gather pollen in quantity it is time for the first colony check.

As is discussed in greater detail below, I recommend that colonies be packed for winter; not everyone in the Southern Tier of New York State does so and even farther north colonies may survive without any special protection. It is likewise important that colonies be unpacked promptly when, or soon after, the first pollen is brought into the hive; in most of the northeast we start to unpack about April 15.

During the first inspection, which is brief, we are concerned with the following items:

1. Dead colonies should be picked up and taken indoors or the entrances plugged so that any remaining honey cannot be removed by robber bees. Dead colonies should be examined closely to determine the cause of death; if no disease is present this honey may be used to feed other colonies. Our chief concern is American foulbrood. If a colony died of a severe infection or nosema or dysentery, however, it would be well

not to use honey from it to feed other colonies. While many persons use and advise the use of drugs to control American foulbrood, I do not. Our area is relatively free of this disease, and I hope it will remain that way; thus, from a practical point of view, we do not need drugs. If colonies die of American foulbrood, precautions should be taken to make certain the disease does not spread, even where drugs are used. It is only by determining why colonies die that one can prevent such losses another year.

2. Live colonies should be checked to determine if they are queenright. One need only see an egg, properly centered in an upright position in the bottom center of a cell, or a larvae or pupae, to be certain a queen is present. It is not necessary to find the queen or to determine how much brood she has or how good her brood pattern might be; this is done at the time of a later inspection. In fact, early inspections can be misleading as brood patterns may be faulty for many reasons early in the year. If a colony has a drone-laying queen (in the case the queen has exhausted her supply of sperm and produces only drones), or if laying workers are present, there is little one can do to save the colony. In the case of a drone layer one may be able to find the queen and kill her and to unite the colony with another queenright colony. This is done by placing the queenright unit on top of the queenless one with a single sheet of newspaper with two or three five to ten inch slits in it between the two units. In the case of a laying worker colony (which is determined by finding multiple and usually small eggs irregularly placed in a cell) the only recourse is to shake the bees out of the colony onto the ground and to place the supers on another colony or in storage. If the supers are placed in storage they should be put onto a strong colony later in the year so they may be cleaned up by the bees.[1]

---

[1]Bees have no trouble cleaning moldy comb or combs containing dead brood providing the brood did not die from American foulbrood; however, this is best done in the warm summer months.

MANAGEMENT FOR COMB HONEY PRODUCTION 55

3. Entrances should be kept reduced and be large enough only to accomodate the bees flying to and from the colony. We usually restrict our colony entrances to a space 3/8 of an inch high by 3 to 5 inches long. Bottomboards should be cleaned (scraped of dead bees and debris). Wet bottomboards should be replaced with dry ones. We always carry a supply of dry bottomboards with us to an out apiary or have some available in an outbuilding. Any other wet or damp parts of a hive should likewise be replaced.

4. Weak colonies should be united, preferably with other weak colonies. The better queen will usually survive. A weak colony is one with less than a pound or a pound and a half of bees (4000-6000 bees). Such a unit will not cover more than a frame when clustered and the amount of brood at this time of the year would be less than four to eight square inches. One must be careful if one judges a weak colony by brood alone. Colonies may have little brood because they may have little pollen and/or honey.

5. In mid-April we hope our colonies will have 30 pounds of reserve honey; not all do and some may have as little as 15 to 20 pounds. I worry if they have less. We usually keep some frames of honey in storage through the winter and give it freely to colonies with little food in the spring. Feeding combs of honey, provided it was produced by disease-free colonies, is the easy way to feed bees. If honey is not available the bees should be fed sugar syrup. In the spring the mixture should be 1:1 by weight or measure; in the fall it should be two parts sugar and one part water.

There are several ways of feeding colonies sugar syrup. We now use one-gallon glass mayonnaise-type jars which we obtain from local restaurants. These jars have screw caps which are four or more inches in diameter; jars with smaller caps tip over too easily. Twenty or more holes, made with a 4d nail, are punched in the cap after the cardboard or plastic liner has been removed. The jar(s) is placed on top of the frames and an empty super put around it. At one time we used tin pails to feed bees but the thin pails made today rust too easily. Some beekeepers use tin pails but coat the inside with hot beeswax or paraffin. Division board feeders are popular with some beekeepers. The plastic, one-piece type leak infrequently and are more practical; however, all plastics have one problem in common and that is they crack easily and have a short life. I don't happen to prefer division board feeders but there is no good reason why they should not be used.

Some people have advocated the feeding of dry cane sugar to bees. The sugar is usually put on top of the innercover and the innercover hole is left open so the bees may have access to the sugar. I've observed some beekeepers pour the sugar onto the top bars of frames in the rear of the hive. In this way the sugar is spread from top to bottom in the back portions of the hive. Dry sugar crystals should not be poured over the center of the brood nest as crystals which fall into cells containing larvae might cause trouble, or at least so it is said.

I think feeding dry sugar is a poor practice. It is true that colonies desperately short of food might survive for a few days on dry sugar. However, all too often I have seen bees carry the dry crystals outdoors and dump them on the ground.

The important consideration as regards food is that between mid-April and the honey flow the bees must not run out of honey. This is especially important in comb honey production where a full complement of bees of all ages and capabilities, especially wax secretors and nectar gatherers (which are of markedly different ages) are needed.

April-May — the first major inspection

Comb honey supers, furniture, sections and foundation cost too much to be placed on colonies which will not fill the sec-

MANAGEMENT FOR COMB HONEY PRODUCTION 57

tions properly. Not all colonies will be used for comb honey production; some will be used to produce the winter stores for those that do. The time of the first full inspection, in addition to checking the items below, is the time to start marking the colonies which are definitely best and will be used later for comb honey production. I don't care to number colonies or to use a dairy. I prefer to keep the top of the inner cover clean and to write abbreviated notes on it. (An accumulation of notes is often interesting to read!) Sometimes I put a 3 by 5 card on the inner cover on which to write these notes; when I do this I usually hold it in place with a thumbtack.

In this inspection we are concerned with the following items: (It is presumed that drone layers, laying worker colonies and those which are too weak will be dealt with as above, or as is described in texts concerned with routine liquid honey production.)

1. Feed as necessary!
2. Give colonies additional room before it is needed by them. Most beekeepers winter their colonies in two supers and under normal circumstances this is to be preferred. A third super should be added sometime between May 1 and 15, depending on colony strength, but definitely ahead of the time the bees will need or even enter it. Only drawn combs should be added at this time; frames with foundation are added only during a honey flow!
3. Check colonies for disease. Sacbrood, European foulbrood and nosema are diseases which are brought about by stress. If they are found the beekeeper should take action accordingly; these diseases are discussed by me and others at length in several books. If American foulbrood is found it should be dealt with according to practices and laws in force in the beekeeper's state.
4. Later in the season, after the comb honey supers are put into place, it will be necessary to make frequent colony in-

20. The first step in examining a colony is to smoke the entrance gently. The beekeeper stands alongside of the colony, not in front of it where he might interefere with flight to and from the entrance.

21. *After the entrance has been smoked the cover is lifted. If the hole in the inner cover is open, as it is in this case, a small amount of smoke is directed at the hole and the bees below.*

22. *The next step in examining the hive is to place the cover, upside down, on the ground to the rear of the colony. When the second super is removed it is placed on the cover. By having the cover rim facing upward there is a surface on which to place the super without crushing bees. The inner cover is lifted with one hand and a small amount of smoke directed over the top bars of the frames beneath. Inner covers may be difficult to remove if propolized heavily; it is not always easy to remove them without prying with a hive tool.*

23. In examining combs in a colony it is best to remove a side comb first since side combs are usually covered with fewer bees and are easier to remove; the first comb taken from the colony is set to one side, but not in a position where it will block colony flight. By leaving one frame out of the super there is additional space for the beekeeper to pry frames apart as subsequent frames are examined and fewer bees will be crushed or killed as manipulations are made. It is very difficult to examine a colony without maiming some bees; the beekeeper should remember that each bee squeezed or crushed is likely to release alarm odor.

spections, usually every seven or eight days. In order to facilitate colony inspections later, this is the time of year to scrape, clean or replace supers, frames and inner covers. I prefer to paint and scrape equipment in the winter and to have a fresh supply on hand to use at this time of the year. Removing propolis and burr comb is easier early in the spring than when the colony is more populous.

5. Check the queen's brood pattern and clip her wings.[2] A good brood pattern is said to be present if the area with brood has few cells without brood and if the brood adjacent to each other is of the same relative age. Capped brood should be adjacent to capped brood; larvae adjacent to larvae, eggs to eggs. When a queen lays in an area it should be apparent that she does not miss many cells.

Young queens usually have better brood patterns than old queens. Young queens are less likely to head colonies which will swarm. Young queens are desired for comb honey produc-

---

[2] Later in the season swarm control will become the beekeeper's chief problem. Despite the fact that we know a great deal about swarming and that there are some reasonably good methods of swarm control, no one has ever been completely successful in preventing it in the production of comb honey. It is for this reason that the queen's wings must be clipped. Clipping a queen's wings will not stop the bees from attempting to swarm, but a swarm whose queen cannot fly will return to the hive. A colony with a queen with clipped wings may try to swarm repeatedly and if left to its own devices, may eventually swarm, taking with it the first virgin queen to emerge from cells which are reared. However, the alert beekeeper will destroy queen cells as they are produced and will prevent this last from happening.

It is easiest to find a queen when the colony population is low. It is for this reason that it is recommended that a queen's wings be clipped at this time of year. Some beekeepers clip the left wing in one year and the right wing of a queen another, thus marking their age for future reference.

It matters little how one clips a queen's wing or wings. I have seen a great variety of techniques described in the literature and I have no doubt that many of them are very good. Some people grasp the queen with their fingers; others catch her under a gauze or disk which holds her in place while her wings are clipped. I prefer to grasp the queen while clipping her wings. I can only suggest that those who would clip a queen's wings might practice on drones. Queens are not inclined to sting. It is important to have a good pair of scissors for clipping wings.

tion. Colonies with poor queens should be requeened. I prefer to requeen by killing the poor queen and placing a nucleus colony with a young queen on top of the one to be requeened. A single sheet of newspaper with two to four slits is placed between the two hives being united. A comb honey producer (any beekeeper for that matter) should have a supply of nucleus colonies, perhaps amounting to as much as ten per cent of his total colony numbers, on hand at all times. Some nucleus colonies should be made with queens bought from the southern states as early as it is practical, usually April 15 to May 1.

May-June — swarm prevention

Several years ago we alerted local police and fire agencies that we wanted and would collect swarms which came to their attention. We have now collected over 200 swarms in a six year period. An interesting picture concerning the time of swarming in this are has emerged: Only two swarms were taken prior to May 15; 78 per cent of the swarms were collected between May 15 and July 15; only two swarms were taken between July 15 and August 15; about 20 per cent of the swarms were collected between August 15 and September 15 and none was found thereafter (see figure 24). Armed with this information we are in a better position to advise when swarm prevention measures should first be undertaken and when swarming is not likely to be a problem, at least in this part of the state. It behooves beekeepers to study the swarming pattern in their area and to implement their swarm prevention measures accordingly.

In this area our first major honey flow starts on July 6 or 7 with the flowering of basswood. On rare occasions bees might make a surplus from clover, but this occurs so infrequently that we do not manage our bees so that they could fill sections even if there were such a flow. Until we put the comb honey

64 COMB HONEY PRODUCTION

supers in place our swarm prevention measures are the same as those which a beekeeper would use to prevent swarming in colonies used for liquid honey production. The following techniques are used:

1. As indicated above, the colonies are supered ahead of the time they will need additional space. If a two-story colony is especially prosperous at the time a third super is added we take one frame of brood from the center of the top super and place it into the middle of the third super. An empty comb from the third super is placed into the middle of the second super. This is called spreading the brood. The presence of brood encourages the queen to move into the third super where there is plenty of space for her to lay. Spreading the brood is an effective way to deter swarming but is also dangerous in that colonies without a sufficient number of bees may not be able to keep all of the brood warm and some or all of it may die. A fourth super may be added, in the same way, two, three or four weeks after the first, depending upon the need for room. Beekeepers with a large number of colonies will often add a fourth super at the time the third is added.

2. At the same time supers are added it is advisable to check between the supers for the presence of queen cups or cells. A number of queen cups may be built two to four weeks before eggs are deposited in them. While the number of cups cannot be taken as a precise measure of whether or not the colony will swarm, it is a good indication of what might take place. The appearance of ten to twenty queen cups, or the addition of fresh (white or light yellow) wax to old cups, is an indication that the colony may swarm if steps are not taken to prevent it.

When queen cups appear in numbers, or even when we find cups with eggs in them, we reverse the supers, thus breaking up the continuity of the brood nest. This is similar to spreading the brood. In reversing, the top super is placed

MANAGEMENT FOR COMB HONEY PRODUCTION

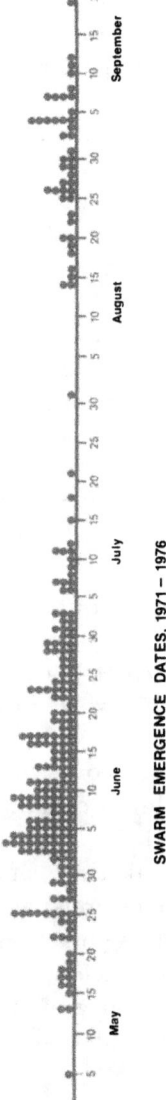

SWARM EMERGENCE DATES, 1971 – 1976

24. *The dates on which over two hundred swarms have been captured in the Ithaca, N.Y. area over a six year period. These data give a beekeeper the information he needs to implement a good swarm prevention program. Further south swarming will occur over a longer period of time, while further north the swarm season will be intensified. Data from the southern states is needed to determine if there is a "no swarm season" during the late spring and early summer in that part of the country.*

on the bottom board and the lower super is placed above. A third super is usually added. Brood may or may not be moved into the third super, depending upon the number of cups and the degree to which the colony is crowded. This method is clearly more drastic than outlined in 1 above; however, it is likewise more effective.

Congestion or crowding in colonies is difficult to define. However, when every frame in a hive is covered with bees, one bee deep, the colony is crowded. Congestion causes swarming.

Colonies should be checked every two weeks during the swarming season and reversed as often as necessary, sometimes as many as three or four times prior to the honey flow. As will be noted below, comb honey producers check their colonies more often after the section supers have been put into place; however, prior to that time every two weeks should be sufficient, provided they are properly reversed and given room each time.

3. Colonies in which capped queen cells, or cells with young larvae, are found, prior to the time the section supers are added, are not suitable for comb honey production. Swarming and queen cell production involve physiological changes in the bees in a colony. Further crowding, as is done when comb supers are added, of a colony which has built cells prior to the honey flow will almost certainly bring about swarming.

Colonies with cells at this stage, and at this time of year, in the management scheme must be artificially swarmed, split, Demareed, etc. All these practices are discussed in texts devoted to liquid honey production; such colonies may be used to produce food for comb honey colonies for winter, or if requeened, may be used for comb honey production in the fall.

Preparation for the honey flow

Colonies should not be reduced to a single super, or sections added, until fresh nectar shakes from the frames and the honey flow has started; however, one cannot delay to long at this stage either.[3] Persons not familiar with the way in which fresh nectar will run out of a comb have an interesting experience in store. When bees are bringing in nectar in great quantity they cannot ripen and remove moisture from it rapidly; at this time if one holds a comb horizontal to the ground, and gives it a slight shake, large droplets of unripe honey will flow from it. At about the same time bees add bits of fresh (white) beeswax to the burr and brace comb already present in the hive. This is called "whitening of the comb" and is also a sign that the flow is in progress.

When we are ready to add sections we first place the deep bottom board, with its rack in place, alongside the prospective comb honey colony. On this is placed a freshly-scraped, deep super of the same color as the bottom super of the colony. The only reason for using a super of the same color is to reduce drifting. We then split the colony into two, three or four parts, depending on the number of supers, placing each of the upper supers on an upsidedown cover. We then search for the queen. This is a time-consuming process and part of the hard work of comb honey production. So far as I am person-

---

[3]Some advocate placing comb honey supers on colonies before the honey flow starts. The reasons for not doing so are several: the colony is prematurely crowded; the bees may chew the wax foundation which may be rebuilt with drone comb which is unsightly; and/or the sections may become travel stained and discolored from excessive deposits of propolis.

An even more important reason for delaying placing the section supers on the hive until after the flow starts is that a honey flow stimulates the wax glands of the workers; wax glands are stimulated and wax is produced anytime there is a honey flow or when bees are fed sugar syrup. Drawing foundation requires a great deal of new wax. Bees may sometimes move wax from one place to another in a hive, and even on a small scale this is not desirable in comb honey production. Bits of old, dark wax, mixed with new white wax, give sections a bad appearance.

68 COMB HONEY PRODUCTION

ally concerned, there is no alternative; I have talked to people who have driven all the bees, and presumably the queen, into the bottom super with a repellent. This method is satisfactory for liquid honey production (when driving the queen down to put excluders in place) but since the method is not always perfect, and sections could be lost if the queen is not in the bottom super, I prefer to make certain by finding her.

When the queen is found she and the frame on which she is located are placed into the empty super which is to become the brood nest for the comb honey producing colony. Following this we place eight more frames in the super, selecting those which have the greatest quantity of capped brood. (Only eight frames are used in total if a follower board is added.) The reason that capped brood is selected is that it will be the first to emerge and thus add more bees to the colony; at the same time more room is provided by the emerg-

25. A poor brood pattern indicating a poor, usually old queen. Some of the cells are empty, some contain larvae and others have capped brood. Good queens have a compact brood nest; eggs are adjacent to eggs, larvae adjacent to larvae of about an equal age and capped brood cells, too are close together. Old queens have poor brood patterns because they fail to lay in every, or almost every available cell, or their eggs fail to hatch.

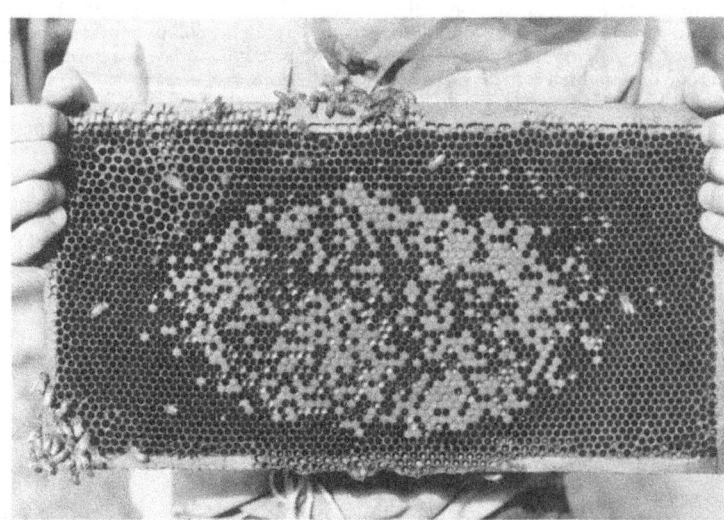

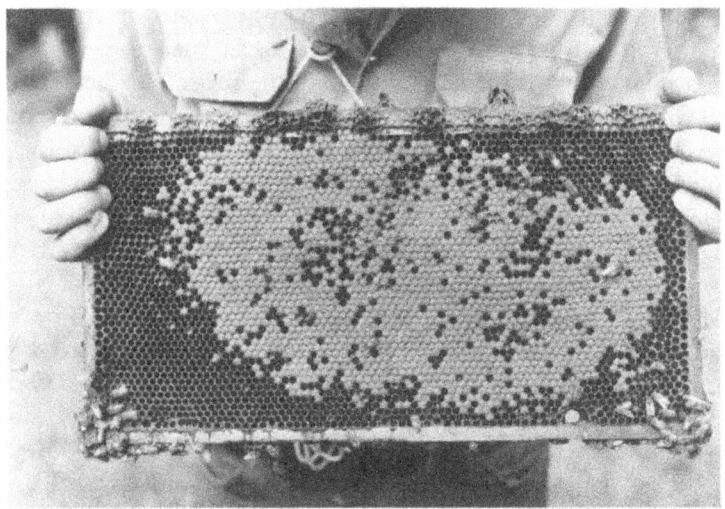

26. *A moderately good brood pattern. Most beekeepers would be pleased to have queens which had brood patterns as good as this one. There are relatively few cells without brood and only a few larvae are mixed with the capped brood.*

27. *An excellent brood pattern; the brood is of almost the same age. The queen cup in the left center is evidence of queen rearing earlier in the year and may remain, unchanged, as it is here, for several months.*

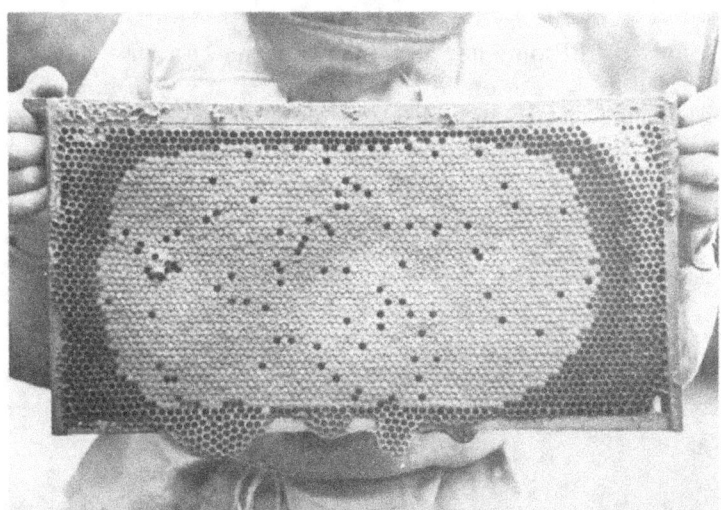

70 COMB HONEY PRODUCTION

ing brood for the queen to lay. The frames are crowded together slightly leaving a somewhat larger than normal bee space between the outside frames and the super. The bees will usually not fill this with burr comb immediately and the extra space will make it easier to remove the frames during subsequent inspections. If the frames had been scraped and cleaned earlier, as was suggested, they will move together without difficulty. Two supers of sections are added; that super with the single bait or starter section[4] is placed immediately on top of the brood nest. The cover and inner cover are added and the colony is checked to make certain there are no unwanted entrances. The extra brood, if there is any, is placed on other colonies. This brood may be used effectively to strengthen newly captured swarms or nucleus (weak) colonies. Colonies strengthened in this manner will often be in a position to produce some surplus honey.

The colony is now ready for honey production. We have also congested the colony and done those things which would encourage it to swarm. Not all comb honey colonies treated in this fashion will build queen cells and attempt to swarm; however, a large number are likely to do so.

Swarm control

When a colony of honey bees swarms between 30 and 70 per cent of the bees and the old queen leave the hive. If a colony is still congested after the primary swarm has left there is likely to be a secondary, and even a tertiary swarm, each of which may leave with a virgin queen soon after they emerge within the colony. Swarming depletes the colony population. Those colonies which swarm are not in a position to fill their

---

[4]Bait sections are discussed in Chapter VII under Starter sections.

28. This nest, built by a swarm which escaped from a commercial apiary, demonstrates the spherical shape of the brood nest in a normal colony. The smallest comb, which is on the outside of the nest and which was adjacent to the boards on the outside of the building, contains the least amount of brood and probably the greatest amount of honey. The second comb has a greater amount of brood. We would expect that the amount of brood in the third and fourth combs would be similar to that in the second comb. A fifth comb, which is not fully visible, would have about as much brood as the first, outside comb.

72 COMB HONEY PRODUCTION

section supers and the beekeeper may be left with half filled sections for another colony to fill.

Why do bees swarm? Dr. C. C. Miller said the following about swarming:

"Upon no other subject connected with bee-keeping have I studied so much, tried so many plans, or made so many failures, as with regard to prevention of swarming. If I knew all about just what makes a colony swarm, I would be in better shape to use preventive measures; but I don't know all about it. Of course I know that want of room and want of ventilation may hasten swarming, and possibly some other things of that kind; but after all there is a good deal of mystery about the whole affair."

As was indicated earlier the queens in colonies which are to be used for comb honey production should have their wings clipped. If this is not done earlier in the season it should be done at the time the colonies are reduced to a single brood nest and the section supers added. Since clipping a queen's wings only deters and does nothing to prevent swarming, it now becomes necessary to make routine inspections of the colony to destroy any queen cells which are present. This should be done every seven or eight days.

When cutting queen cells I prefer to shake the bees from the frames. I usually first remove the follower board or a frame on the side of the brood nest and set it outside of the hive. As subsequent frames are manipulated the bees are shaken from them but within the colony. In the event that the bees are shaken outside of the hive there is a danger that the queen might be lost in the grass. It is correct that queens cling to combs more tenaciously than do worker bees and oftentimes after having shaken the bees from a frame I have found the queen still on it.

I have noted that men who produce comb honey usually have better frames in their brood nest than do those con-

MANAGEMENT FOR COMB HONEY PRODUCTION

cerned with liquid honey production. I have also noted that comb honey producers do not care to use frames with split bottombars. Honey bees don't think and they don't hide queen cells with a vengeance for the purpose of the cells not being found by the beekeeper; however, it often appears as though they are doing so! It is said one may find queen cells between the bars of a split bottombar if the bars are warped out of shape. One may also find queen cells in the corners of combs and sometimes partially hidden by burr and brace comb. It is for the reason that there is much less opportunity for the beekeeper to miss queen cells which should be cut and removed that most men prefer to have very good combs in their colonies' brood nests.

After one has cut queen cells from a colony two or three times (and has made a note of the fact on the hive innercover or some similar place) it is apparent that the colony is determined to swarm and the beekeeper often feels that more drastic measures should be taken to prevent swarming. Dr. C. C. Miller and others investigated both removing the queen and caging the queen as swarm control measures. In both cases the colonies would be inspected for queen cells (which would be removed if present) before the queen was released or replaced. The technique worked well sometimes but not always. It may very well work better in the hands of some beekeepers than in others; caging or removing the queen may also be more effective during certain seasons. I do not care to recommend either method as a swarm control measure.

Still another option for dealing with a colony which persists in building cells is to remove the comb honey supers, place them on other colonies, and to use the colony for liquid honey production. This may sound like "giving up" but oftentimes too much time may otherwise be wasted with a colony which may not produce much honey in any event. Colonies which persist in queen rearing in this way should be marked for requeening.

74 COMB HONEY PRODUCTION

Late summer and fall management

The date for removing the last comb honey supers will vary from one area to another. For example, in areas where there is both clover and alfalfa, it is entirely possible that the producing season may extend well into August. In much of the northeast, alfalfa yields nectar in great quantity on the second and third hay cuttings; everywhere it is produced alfalfa honey is of high quality. Alfalfa honey is usually sold as clover honey, which it resembles. Many honey producing areas are one-crop areas, areas which produce only clover honey. In other parts of the country the white clover flow may fade into or be followed by a goldenrod flow. Goldenrod may yield nectar anytime in August or September. It may take three to five years for a beekeeper to assess a territory completely and to arrange his management schedule to his satisfaction.

In any event, when the last comb honey supers are removed from the colony it is time to consider those steps one should take to produce a maximum crop the following year. Annual requeening of colonies has been advocated by many persons and for good reasons. Still, annual requeening is not popular because of the time and cost involved. There are good data to show that colonies with queens less than a year old are much less likely to swarm than those with queens more than a year old; that fact has been mentioned earlier but cannot be emphasized too much.

Insofar as comb honey production is concerned, I recommend annual requeening of colonies. This may be done in a great variety of ways but requeening the colony in August after the last comb honey supers have been removed is probably the easiest. Furthermore, if the colony is not successfully requeened the first time there will be no honey lost and there is still an opportunity to attempt requeening again before the colony is prepared for winter.

MANAGEMENT FOR COMB HONEY PRODUCTION 75

There is no perfect way to requeen a colony. A great number of methods of requeening have been described but the one which meets with the greatest success involves combining the colony to be requeened with a nucleus colony. The nucleus colony is placed on top of the dequeened colony with a single piece of newspaper between as has been discussed above.

September — the food chamber

It is probably easiest to winter colonies in two supers with approximately 60 to 80 pounds of honey. Many beekeepers in the North winter their bees in three supers with a similar amount of honey but I think using the third super is unnecessary and extravagant. Several decades ago it was proper to winter bees in single supers, feeding them sugar syrup in the late fall for winter food. The method worked well but was very time consuming, and has largely been abandoned because of the labor involved. This has been discussed in chapter I.

In most parts of the country colonies of honey bees which have been used to produce comb honey will not have sufficient honey for winter. In areas where there is ample goldenrod, and the beekeeper does not take advantage of the flow to produce more comb honey, it may be possible for the bees to store sufficient food for winter. Goldenrod honey is probably not so good for winter food as is light clover honey because it contains more indigestible material and there is greater accumulation of feces in the bees. However, we have wintered many colonies successfully on goldenrod honey and there is no reason why it should not be used for winter food.

In single honey flow areas it is necessary to use a given number of colonies to produce the winter food for others. Depending upon the time of year that comb honey supers

76 COMB HONEY PRODUCTION

are removed, it is satisfactory to place the full supers of winter food on the producing colonies in late September.[5]

October-November — packing

I recommend that bees wintered in New York State and those states with a similar latitude, be given some form of winter protection. There have been several studies on wintering. The consensus is that those colonies which are given an outer wrapping of paper, provided proper ventilation and some type of packing on top, will have more bees and brood the following spring than those colonies which are not given protection. However, the question is debatable and I am aware of the fact that several successful beekeepers winter their bees successfully without any special preparation.[6]

---

[5]Races of honey bees vary greatly as concerns brood rearing at the end of the honey flow. Some races of honey bees will cease brood rearing or slow it considerably immediately at the end of the honey flow. Other races are known to continue to rear brood so long as honey is available to them. We have no pure races of honey bees in the United States though bees which are predominately of the Italian race are most common. Because there has been a great mixing of bees in the United States, colonies, even in the same apiary, may vary as regards brood rearing and the amount of honey available to them. Therefore, if the honey flow stops as early as the end of July it may be desirable to store the winter food in a honey house and to place it on the colonies in late September or early October. This will still allow the bees sufficient time to arrange their nest for winter.

[6]For details on a method of wintering honey bees see Information Bulletin 109, "Wintering Honey Bees in the Northeast", by E.J. Dyce and R.A. Morse, available for 50¢ by writing: Mailing Room, Research Park, College of Agriculture and Life Sciences, Cornell University, Ithaca, N.Y. 14853.

V.
A Special Management Technique

Raymond Churchill of Watertown is one of our more successful comb honey producers in New York State. He says the technique outlined below is not his original with him but he has used it successfully for many years. Ray has taken more prizes for his comb honey at Eastern Apicultural Society honey shows than any other single beekeeper.[1] In some ways the method resembles the "forced swarming" system which Dr. C. C. Miller discusses. Churchill's early spring management follows that already discussed.

One or two days after the honey flow has started the colonies are still in three or four supers. A colony to be used to produce comb honey is broken apart, that is, the upper two or three

---

[1]Witness to Ray Churchill's ability as a producer of comb honey is the fact that he won first place, and a silver trophy, at the Eastern Apicultural Society honey show again in 1977. That year stands in history as one of the poorer production years for New York State, especially in northern New York and the Watertown area which has a long history of being a location where colonies produce good crops of high quality honey. How did Ray do it? In addition to his usual home yard, where most of his comb honey is produced, Ray had a single colony in a location remote from all other colonies. Bees from that colony were not forced to share the forage in the immediate area with bees from other colonies. And, in 1977 when forage was in short supply, the effort paid off, for it was from that single colony that Ray harvested the three winning sections. (Three sections are required for a single entry in the E.A.S. show.) Does Ray practice that routine every year? I don't know — but I do know that it takes expertise and effort to be a consistent winner at honey shows!

78 COMB HONEY PRODUCTION

supers are set off individually on upsidedown covers. Separating the supers in this fashion facilitates finding the queen. The bottom super and the bottom board are set to one side. A new (clean) bottomboard and a half depth super, with only one half depth drawn comb containing some brood, is placed on the original colony's stand. Nine more half depth frames, with foundation, are placed into this half depth super.

The next step is to find the queen, clip her wings, if this has not already been done, and to place her on the brood frame in the half depth super. Two comb honey supers, without bait sections, are placed on top of the half depth super. The inner cover and cover are put into place.

Next comes the hard work of shaking the bees from the three or four supers, frame by frame. The bees are shaken immediately in front of the new supers on the old location. It is best to sort the frames as this is done; the combs containing brood are placed on a weak colony which is to be strengthened. (Since this manipulation is usually done in July there is not too much danger of the brood being chilled.) Those combs which are empty, or which contain honey, are placed on other colonies.

It is well to emphasize that one should wait until the honey flow has been in progress for a day or two before reducing the colony to one super and putting on the section supers. Two days are preferred so the worker bees' wax glands will have been stimulated and the bees are in a physiological condition to secrete all the wax which is required. In our experience nothing stimulates wax production as does a honey flow; still, time is required for glandular development to take place and for wax to be produced.

Reversing supers

The normal procedure used in reversing comb honey supers is not used with the Churchill method. The first time I

A SPECIAL MANAGEMENT TECHNIQUE 79

tried this method the queen laid eggs in seven sections of the bottom comb honey super. When I complained about this to Ray he was quick to point out that I had not paid sufficient attention to his instructions; it is necessary to reverse the position of the comb honey supers when the cells in the bottom comb honey super are about one third drawn. (The cells in the bottom super will be drawn before those in the top super.) If this is not done the queen may move into the section supers and lay eggs. She is encouraged to do so both because of the restricted brood nest and the fact that the top bars of the half depth frames are usually less thick than those of full depth frames. Thick top bars, spaced close together, tend to act as a queen excluder as is discussed in chapter 1. It is possible to buy or make half depth frames with thick top bars and this is to be preferred when this technique for making comb honey is used.

As additional supers are given to the comb honey producing colonies they are added below those already on the colony; however, they must be rotated out of that position within a few days and a super with nearly filled sections put into the lower position. This is the only way in which one may prevent the queen from laying eggs in the sections. In this regard there is more manipulation of supers than takes place normally in producing comb honey.

Swarm prevention

Churchill normally puts his comb honey supers on his colonies about July 1. This is during the latter part of the swarming season. This is one reason why his colonies are not inclined to swarm. More important, however, is the fact that nearly all of the brood is removed. Despite the fact that the colonies are crowded with bees they are not inclined to swarm because they have no brood.

I suggest that clipping the queen's wings is advisable, even when using this method of comb honey production. It is an added precaution which will not prevent swarming but which will certainly thwart the process by several days.

Preparation for the following year

Colonies of honey bees reduced to a half depth super may or may not be saved for the following season. In the southern states such a unit would have little difficulty surviving following the primary honey flow. In the North their survival depends upon the length of the season following the major flow and whether or not the colony has time to gather its winter stores. It is, of course, possible to feed a colony either by placing a super of honey on it or by using sugar syrup.

If one is interested in keeping the same number of colonies from year to year, the simplest technique is to make a new colony from the brood taken away from the producing colony; the brood is added to a nucleus colony. The possibilities are many; what is done depends on the location and its potential and the number of colonies the beekeeper proposes to manage.

VI.
Cut Comb Honey

In this chapter we are concerned with combs of honey, usually produced in half depth frames, which are used to make chunks of honey wrapped in plastic and sold like comb honey, or placed into jars and surrounded with liquid honey. Producers of cut comb honey make a product as delicious as comb honey. A perfectionist might say it is a poor substitute for comb honey but this is not precisely so. In defense of the perfectionist it can be said that the man who produces comb honey successfully, year after year, may be the best manager of bees.

Anytime one crowds or congests a honey bee colony the swarming instinct is encouraged to show itself. One must crowd a colony to produce cut comb honey or comb honey. There is better ventilation in a super with cut comb honey frames than exists in a super with sections; still, the colony is more crowded than when it is being used to produce liquid honey; under these circumstances swarm control measures, and rather exact ones, must be exercised.

Most cut comb honey is made in frames which are 4 1/4" or 5 3/8" deep. The wooden frame is reused year after year.

Cut comb honey is predominately a southern product and

probably originated in Georgia. Packing cut comb honey in jars and surrounding it with liquid honey is also a southern innovation. It has been commercially feasible in the south because most honeys which are produced in many of the southern states, especially Florida and Georgia, tend to granulate much less rapidly than do honeys produced in the northern states. To most people a partially granulated, chunk-liquid honey pack is unattractive.

The shelf-life of a chunk-liquid honey pack is limited (because of the danger of crystallization). Because this is true, and it is not commercially (financially) reasonable to recover the honey from such a pack, some large honey companies have refused to pack jars of chunk honey surrounded with liquid honey. One cannot liquify the honey in a crystallized chunk honey pack without melting the wax. However, a chunk honey pack is an attractive item and beekeepers with only limited sales and who pack their honey frequently will find it a good sales item.

Management for cut comb versus comb honey

Producing cut comb honey has two elements in its favor. The first is that it is easier to prepare cut comb frames than it is to prepare sections. Secondly, there is much better ventilation in a colony producing cut comb honey and the tendency to swarm is reduced by a small percentage. However, to insure that every comb is filled to its maximum capacity it is usually necessary to reduce the colonies to a single brood nest as is done when producing comb honey. This causes congestion of the brood nest and the same swarm control measures which are exercised for producing comb honey must be used in the production of cut comb honey.

A disadvantage in producing cut comb honey is that there is less furniture in the cut comb super and the queen is more

CUT COMB HONEY

inclined to move into a super and lay eggs. Occasionally producers of cut comb honey resort to the use of excluders, though this is rare; they should not be necessary. Clearly, one could not reduce a colony used to produce cut comb honey to a single half depth super as is outlined in Chapter V.

All other factors are the same as when producing comb honey. Colonies should not be reduced to a single brood nest until the honey flow has started as wax must be produced in great quantity as it is in the case of comb honey. The hives should be level. There should be no upper entrances. Supering colonies is essentially the same as when producing comb honey. Rotating the bottom super earlier may be necessary to deter the queen's moving into it and laying eggs.

Millard Coggshall, who managed several hundred colonies of bees in central Florida for cut comb honey for many years, wrote the following when I asked him about excluders for that area: "From my experience, a queen excluder seems to reduce the willingness of a colony to produce cut comb honey. I produced cut comb on top of a one and half story colony. About two to three inches of capped honey in the top half story not only acted as a good queen excluder, but also kept the bees from storing pollen in the cut comb honey. The pollen is more trouble than the occasional patch of brood, which could be cut out. Cells of pollen scattered around in the cut comb honey make the comb unusable."

Savings in cut comb honey production

Other than saving the obviously high cost of sections there are other savings to be made when cut comb honey is produced. One does not have the expensive furniture to purchase which is used in the case of comb honey. Cleaning the cut comb honey frames is not so difficult as is cleaning section furniture. The very thin comb foundation which is used to make comb honey is the same as is used to make cut comb honey.

Spacing the frames used to make cut comb honey is obviously a delicate matter. The proper spacing must be observed or the bees will build brace and burr comb between the frames. It is for this reason that many beekeepers prefer to make homemade, self-spacing frames for chunk honey.[1] Whether one uses homemade frames or factory-made frames does not matter but it is probably best to use only nine frames per hive. Some beekeepers use only eight. The extra spaces on the sides of the super are filled with blanks (dummy boards) usually made from thick lumber. The purpose of using fewer frames, as in the case of producing comb honey, is to have the foundation which is to be drawn immediately above the brood nest. Bees do not do so good a job of drawing foundation and filling cells in the outside frames.

Toward the end of the flow it may be helpful to rotate frames within a super, moving those from an outside position to the center. As is indicated, the same may be done with sections but it is much more difficult to move sections than it is to move cut comb honey frames.

Still another saving in the production of cut comb honey is that those combs or pieces of comb not completely filled might be crushed and the honey removed from them. In this way both the wax and honey may be saved and there is not the additional loss of the wooden sections as takes place when one recovers the honey from partially filled comb honey sections.

Preparation of the frames

Most producers of chunk honey use full sheets of foundation so as to have a uniform comb made of worker cells. As has been indicated, a frame of capped honey with worker cells

---

[1]Self-spacing is accomplished by controlling width of the frame's end bars.

CUT COMB HONEY 85

has a more attractive appearance than one with drone cells. Usually the foundation is fixed into the frame with hot beeswax. Paraffin may be used. Since no wire is used in the frame, the foundation cannot be affixed on the endbars. It is best to have a suitable framework or mounting board when putting the foundation in place.

As is the case in mounting any foundation in a frame or section, there is great danger of its warping and buckling as a result of changes in temperature. The problems of foundation buckling are discussed in Chapter II, When to put foundation into sections.

Cleaning the frames

When one makes only a few frames of cut comb honey it is satisfactory to scrape and clean the frames as one would the normal furniture in a comb honey super. The slot in the underside of the frame's topbar is cleaned using a nail or the sharp end of a hive tool.

Beekeepers who produce cut comb honey on a large scale usually place the supers, with the frames in place, in boiling lye water for one to two minutes. The lye dissolves the wax and propolis thus cleaning the frame. These supers are then dipped into a hot water bath with fresh water or hosed with hot water to remove any remnants of the lye. Since there is usually very little propolis in the case of cut comb honey frames, a dipping of one to two minutes in the lye in satisfactory.[2] The removal of the propolis and wax on a cut comb

---

[2]Beekeepers often use boiling lye water to clean old frames from which the comb has been removed or in instances where the frames have been broken in some way. Again, the method is satisfactory but old frames must usually be soaked from 5 to 10 minutes and this has some adverse effect on the wood, making more extensive renailing necessary.

Boiling lye water can cause serious skin burns and should be used with great care. In making a lye solution the lye is added when the water is cold and then the solution is brought to a boil. Pouring lye into boiling water will have an explosion-like effect.

frame may weaken the corners and help to loosen a few nails. However, the renailing problem is not serious and there is little problem with wood deterioration.

Cutting cut comb honey

There are a variety of ways to cut and prepare cut comb honey for market. The actual cutting is made easier by using a hot knife or instrument. On a small scale, one may cut comb honey, one piece at a time, but gadgets for cutting a whole comb into pieces of varying sizes have been developed.

Whether one proposes to wrap individual pieces of cut comb honey or to place the pieces in jars surrounded by liquid honey, the pieces of cut comb should first be drained. In the case of individual packs one does not want a glob of liquid honey underlying the piece of comb honey. In the case where the cut comb is surrounded by liquid honey, granulation may be deterred if the liquid honey from a piece of cut comb is first removed. In most instances the liquid honey used to surround a piece of cut comb honey is pasteurized so as to deter both fermentation and granulation.

There are two methods of draining pieces of cut comb honey. One is to place the pieces on pans or trays in a warm room and allow them to drain for 12 to 24 hours. Another technique is placing the pieces in an extractor and revolving it at a very low speed so as to remove the liquid honey. In this case pieces of cut comb are usually placed on half-inch hardware cloth so that the liquid honey may flow freely from the pieces of cut comb. It is important to revolve the extractor sufficiently slow that the outline of the half-inch hardware cloth is not imprinted on the pieces of cut comb honey.

Packaging cut comb honey

There are a great number of ways of packaging cut comb

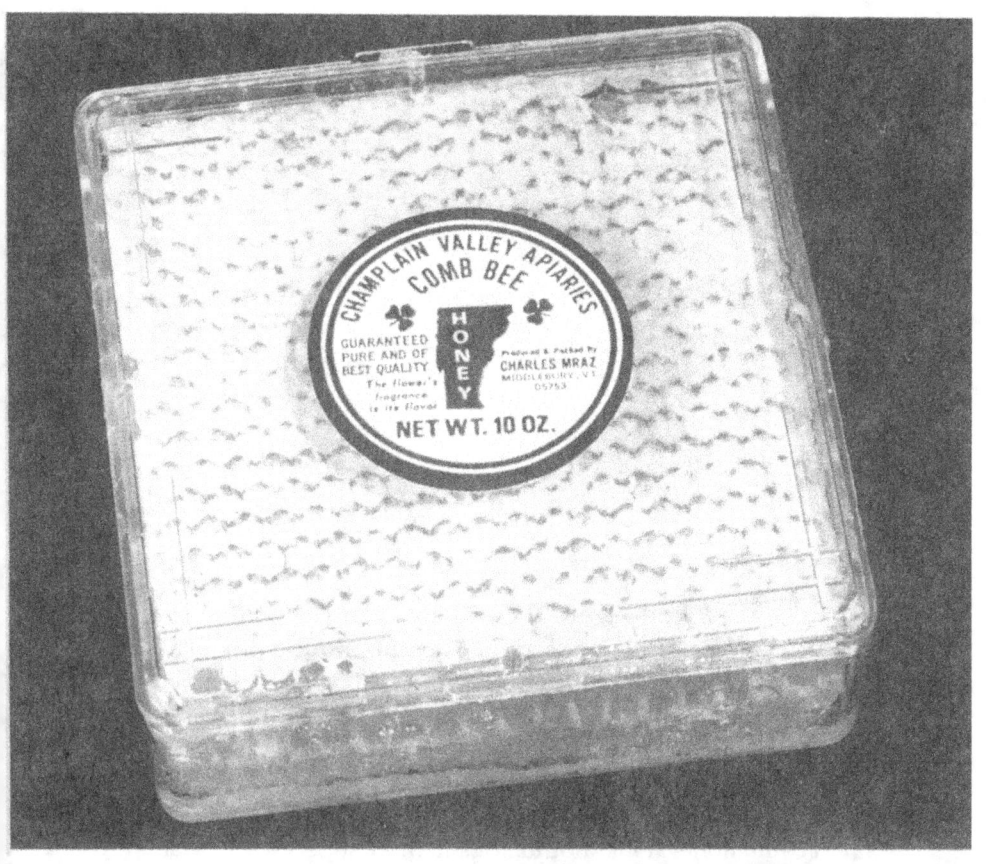

29. *Cut comb honey in a plastic tray; the hard plastic makes an attractive package and serves to protect the comb much better than does cellophane.*

honey. Most beekeepers place the pieces of cut comb in plastic or cellophane envelopes which are heat sealed to prevent leakage. Pieces of cut comb honey, like comb honey itself, are often pinched, squeezed, and otherwise damaged by the interested, and sometimes buying, public. To protect their product some beekeepers have had heavy cardboard or plastic tubs made in which the cut comb pieces rest.

VII.
Supering

Part of the art of successful comb honey production involves adding supers, called supering. Supering, in the production of liquid honey, is simple; one merely adds supers, one above the other, usually giving the colony one or two more supers than it would usually need in the event of an especially strong honey flow.

The matter is not so simple in the case of comb honey or cut comb honey. One should not add more supers than the bees will begin to work in immediately. Bees may chew foundation or add bits of propolis or otherwise stain sections in which they are not drawing wax and all of this may make the sections unsaleable. Specific aspects of supering are discussed below.

Supering for comb honey

Adding supers in comb honey production is a precise science which was studied at length many decades ago. Supering correctly is a factor in swarm control and gives better control over the storing instinct. One should not give too much room at the beginning of the honey flow or the bees may spread their efforts over too great an area. If, however, one gives too little room there may be idle bees, especially comb builders. As supers of comb honey are filled bees are crowded out of them; if the supers are not added in the correct order the bees may be crowded out of the supers and into the brood nest. This could encourage swarming.

Since there must be some comb building before bees can store any nectar, and comb builders are young bees, new (empty) supers are added under those already on the hive. This is called bottom supering. At the same time this gives the greatest amount of space immediately above the brood nest where it is most needed. Placing foundation above the brood nest is also a deterrent to the queen's laying in sections; while queens usually do not do so, it is not impossible that they may lay eggs in sections when the brood nest is crowded.

At the same time the super which contains the fullest sections should be kept immediately above the most recently added super. In this way it is in close proximity to both wax-secreting and nectar-gathering bees; thus, the sections in it will be finished more rapidly and may be removed quickly and before they are travel stained excessively. If, on the other hand, one does not keep the fullest super in the correct position the sections may not be finished properly and may lose their fine appearance for market because of travel stain.

The number of supers which are added depends entirely on the intensity and length of the honey flow. Honey flows from basswood, for example, are of short duration, usually not more than 10 to 12 days. In such a honey flow one does not have time to add or use more than the one (sometimes two) super which is placed on the colony in the beginning. Clover honey flows last longer and a fortunate beekeeper may add three or four supers in addition to the one or two first given the colony. It may take several years of experience before an area is understood sufficiently well that errors in supering are not made.

90 COMB HONEY PRODUCTION

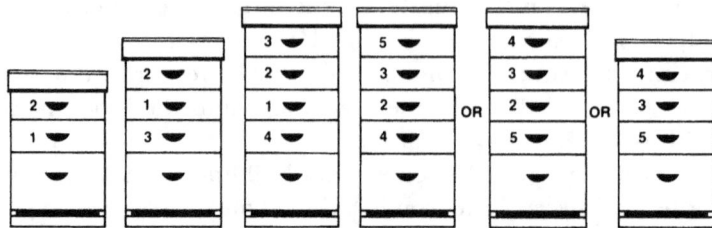

30. *Demuth's system of supering for comb honey production, modified. This method is valid only in areas, and years, when colonies may be expected to fill four or more supers; by contrast, producers of extracted honey would expect to produce a minimum of 100 pounds of honey and more likely 125 to 150 pounds per colony under the same conditions. Obviously not all beekeepers will be so fortunate as to live in an area where the honey flow is this good or production per colony is so high. Many comb honey producers must be content to produce one or two supers of comb honey per colony. See text for the reasoning behind adding supers in this manner.*

Figure 30 is modified from that illustrated in George S. Demuth's 1919 bulletin on comb honey production. His method takes into account all the considerations outlined above. Demuth added only one super at the start of the honey flow. I believe that it is reasonable, during most honey flows, to add two supers at the outset. Care must be exercised near the end of the flow so that colonies are not oversupered; for this reason, three variations in the final supering are shown.

Leveling colonies

Colonies used to produce comb honey should be level. It is ordinarily recommended that colonies tip forward slightly, and with good reason. During most of the year tilting a colony

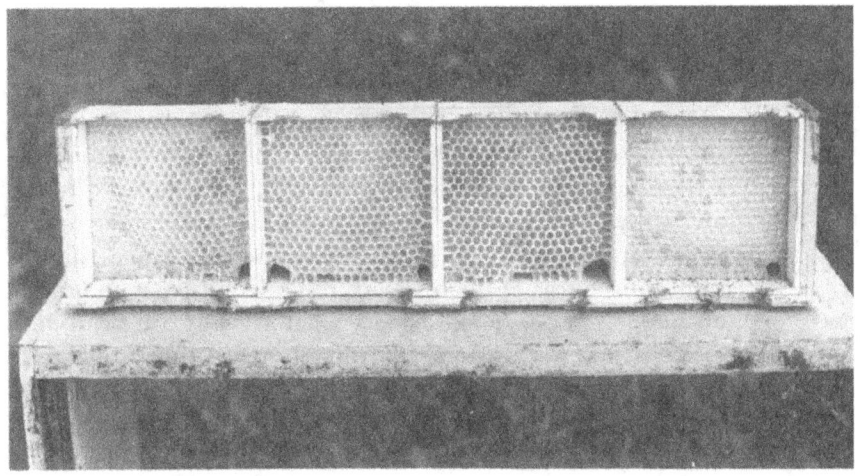

31. These partially-filled sections illustrate the natural chimney effect in comb building which occurs in a honey bee colony. The two inside sections are drawn about equally. The outside section to the left was to the rear of the hive while that to the right was in front; it is natural for bees to draw that foundation over the brood nest first and there is a tendency for them to draw the foundation to the rear of the hive before that in front is drawn. It is for this reason that comb honey supers are turned end for end during the honey flow.

32. Partially-filled, square sections showing the chimney effect. This effect causes no problems during good honey flows; however, in poor flows, or at the end of the season, the beekeeper may be left with a number of unmarketable sections such as this one.

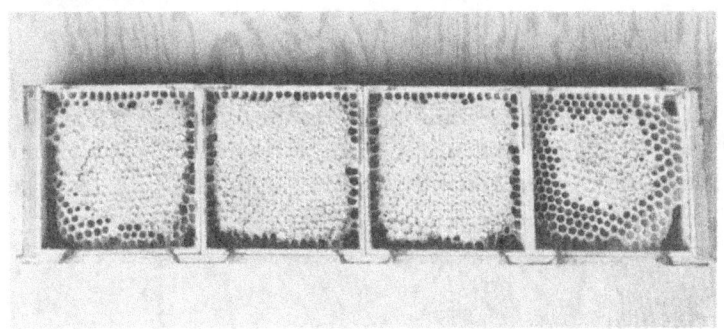

92 COMB HONEY PRODUCTION

so the entrance is protected by an overhanging cover is good practice: the alighting board is kept dry, or free of ice and snow; also, should rain be driven into the entrance because of a high wind, the water drains from the bottomboard quickly and with ease. However, if a colony is tilted while the honey flow is in progress, that portion of the storing frames, or sections, at the rear of the hive will be drawn, filled with nectar, and capped first. There occurs what is termed a "chimney effect": the combs centered directly over the brood nest receive first attention. The problem is eliminated by leveling the producing colonies at the time the section supers are added.

Turning comb honey supers end for end

Leveling colonies does not always ensure proper filling of sections from one end of the section holder to the other. Two factors are involved: if the frames in the brood nest below the section have more brood near one end of the colony, then it it likely that the bees will do a better job of filling the sections immediately above the brood. Secondly, there does seem to be a tendency for bees to store honey toward the rear of the colony even when it is perfectly level. This may very well have to do with ventilation in the hive. For these reasons it is often advisable to turn comb honey supers end for end which will eliminate the problem. This is usually done when the section supers are two-thirds to three-quarters filled.

Closing upper entrances

Colonies used for comb honey production should have no upper entrances. When upper entrances exist, the comb near the entrance(s) will be travel stained and travel stain is to be avoided as much as possible. Also, and perhaps more important, bees are reluctant to store honey near an entrance. It will be noted, for example, in a bee tree, or when bees are removed from the side of a house, that the brood nest is near the entrance and the honey is stored some distance from it. Sections in the vicinity of an upper entrance will not be properly filled.

The fact that upper entrances are not used in comb honey production reduces the opportunities for ventilation of the colony. It is just another reason why swarm control is more difficult but the problem cannot be avoided. Another reason for advocating deep bottomboards, as was done in chapter I, is to improve ventilation and in part to circumvent the problems caused by not allowing upper entrances during the producing season.

Starter sections

Starter or bait sections are those which are partially drawn and are used to attract or draw bees into supers. Most comb honey producers use them religiously. Usually only one section is used per super and then only in the first super which is placed on the colony. The starter section is placed near the center of the section super.

From a biological point of view I have often wondered if starter sections were necessary. I am inclined to accept management methods which are used by successful beekeepers even though the thinking which has gone into the system is not always clear. Comb honey producers with whom I have talked about this say that bees will enter the comb honey supers faster (and therefore go to work more rapidly) if starter sections are present. I am inclined, in this case, to accept what they say as fact. One explanation may be odor. We know that odor plays an important role in the life of the honey bee. It

94 COMB HONEY PRODUCTION

is possible that the odor of the new section wood and foundation may be somewhat repellent to bees; if this is so, it is possible that bait sections are very important. In any event, I recommend their use.

Most beekeepers use as starter sections those which were not completely filled the year before and which they allowed the bees to rob out. Comb in such sections is tough because additional wax has been added by the bees in the process of reworking and rebuilding the comb. Starter sections, even though they may have a good appearance, should not be sold. Persons buying comb honey expect only the best. The tops of the sections should be marked with paint or a penciled X so that they may be easily distinguished and sorted out from the rest.

One of our successful New York State comb honey producers, who until he retired made his entire living with bees, made his bait sections a little differently. In the middle of the honey flow he removed from good producing colonies several supers with sections in which the cells were drawn out only one eighth to one quarter inch. These supers had been on the hives only two or three days. At this stage the bees have not yet put honey in the cells. The supers were wrapped in paper so as to prevent the sections becoming dusty. The sections were left in the super and not taken out until they were needed as bait sections the following year. This man told me the wax in such sections is not tough and when filled with honey that they could be sold as normal sections. He assured me he had checked this; however, I have never tasted such a section myself. I think he was probably correct; he was on other management techniques we discussed!

Go-backs

Go-backs are those sections in an otherwise finished super

of sections which must be returned to the bees to be filled. They may be sections 80 to 90 percent full but with enough empty cells to be unsightly. They are the bane of the comb honey producers. Go-backs may be eliminated often, or their number reduced, by turning supers end for end as is mentioned above, giving fewer supers, centering the sections above the brood nest with greater care, having more populous colonies, or through the use of other routine management techniques.

Not all go-backs are the result of poor management. Partially filled sections occur when a honey flow stops abruptly; there is little a beekeeper can do about that. At this point it is well to emphasize again that not all locations are good for comb honey production. Persons living in areas with intermitent, undependable honey flows must content themselves with the production of cut comb or liquid honey or be willing to suffer the financial loss which accompanies discarding partially filled sections.

The best method of handling go-backs is to sort and return them to colonies for finishing promptly. This may be done in the apiary but is better done in the honey house or shop. There is no question too that some colonies do a better job of finishing sections than do others. Some colonies cap cells more rapidly and may also use less propolis and have whiter cappings. As section supers are removed the better producing colonies should be noted and used to finish the undone sections.

Feeding colonies honey to fill partially filled sections after the honey flow has ceased is a subject almost fastidiously avoided by writers on comb honey production. This is probably true for two reasons: there almost appears to be something artificial about it, and more important, it is a difficult and sometimes an impossible task to finish sections by doing so.

Diluting extracted honey with water and feeding it from

96 COMB HONEY PRODUCTION

pails or jars placed above the sections is a poor method of filling sections. Colonies fed in this way must be populous, have the brood nest super already filled with brood and honey, and be fed a large volume of diluted honey rapidly. The frequency of success is such that I do not suggest one's trying the method except on a trial basis. Sections successfully filled using this technique are frequently travel stained excessively. Most of the honey fed in this manner is used to grow more bees.

An interesting method of feeding bees for winter which may be sometimes adapted to fill unfinished sections involves placing filled combs of honey under the brood nest in the late fall. The system will work only in mid or late October in the southern tier of New York State and probably similarly late in other parts of the country. If four or five frames of capped honey are spaced equally in a ten frame super, and if the cappings are raked and broken (such as with a hive tool), the bees will move the honey above the brood nest and clustering area. Why bees will move honey only in the late fall in this manner we do not understand. They will not remove honey from cells which have unbroken cappings. One can demonstrate this phenomenon by writing their initials or a design on a comb by carefully breaking the cappings on only certain cells. The disadvantage of this technique to fill go-backs is that one must depend on a suitable season. More important, the colony must be in a single brood nest and may be left with insufficient stores for winter. Beekeepers who practice this technique, and they are few, add a full super of honey to the colony after the section supers are removed and hope for the best. Again, travel stain may be a problem.

Beekeepers with extensive holdings often have an apiary which may have a late honey flow which is consistent. This may result from special plantings in the vicinity. If the color and flavor of the honey is similar to that produced earlier, such flows may be used to fill sections. However, it is evident that one mentions the above possibilities almost in desperation; this only emphasizes that comb honey producers must be alert to every system to succeed.

In dealing with go-backs, we emphasize that practicing those management techniques which reduce their numbers is the preferred route. This is the fulfillment of the art and science of keeping bees successfully!

VIII.
Removing Comb Honey Supers

Individual comb honey supers should be removed as soon as the sections in them are finished. The chief reason, of course, is to avoid travel stain, a problem which cannot be overemphasized. Unlike the producer of liquid honey who waits until the end of the honey flow to visit his apiaries, the comb honey producer visits and manipulates his colonies frequently; there is no alternative if one is to produce first class sections.

A lesser problem than travel stain is that bees may open some cells and remove honey from them at the end of the flow. Some races of bees cease or slow brood rearing if a honey flow is interrupted or stops; others do not but continue to rear brood so long as food, honey and pollen, are available to them. Since the bees we use in the United States are a conglomerate, a mixture of many races, from many parts of the world, one is never certain when or where a trait is likely to occur. Queen breeders usually advertise their stock as being one race or strain or another; however, to be perfectly honest they should indicate their bees are predominantly this or that.

When removing comb honey sections it is important that all the bees be removed from the super before they are carried from the apiary. Even one or two bees may uncap a number of cells in a short period of time and thus ruin many sections. When fumigants were still legal it was common practice for comb honey producers to fumigate their section supers in the apiary immediately after they were removed from the colonies. This gave protection both against straggler bees and wax moths.

Smoke particles and the danger of oversmoking

It is normally impossible to open and examine a colony of bees without the use of smoke. When removing supers with a repellent or a bee blower, it is standard procedure to "start the bees down" by blowing smoke over the super. Smoke calms bees but it also has a repellent effect. In the case of liquid honey the removal of supers may be speeded up by the careful use of smoke.

In the case of removing comb honey, care in the use of smoke must be exercised for two reasons: Most smoker fuel disintegrates leaving a small amount of ash. Much of the fine, light weight ash is discharged from the nozzle of the smoker as it is used and the particles are lodged on the combs and in the hive. The beekeeper seldom notices these particles because they are carried from the hive by the bees. When comb honey supers are removed the black specks may remain on the white comb surface and ruin the appearance of the sections. Smoke particles on comb surfaces may be avoided by using less smoke. A helpful technique is to recharge the smoker more often using burlap or twine which may also act as a filter so long as there is some unburned material on the top of the smoker. Another popular method is to cover the hot fuel with a handful of green grass which serves to cool the smoke and act as a filter; however, grass dries rapidly in a smoker and more grass must be added at frequent intervals.

A more important reason for the careful use of smoke is that most smoked bees immediately engorge with honey. This

COMB HONEY PRODUCTION

might pose no problem if the bees took the honey from uncapped cells. The comb honey producer works hard to have all the cells in his sections capped and usually few such cells are available to the bees. Oversmoked bees are capable of uncapping cells, or portions of cells, rapidly. Many comb honey sections have been ruined by beekeepers who use too much smoke.

Robbing

Perhaps robbing is to simple a problem to mention. However, in the case of comb honey a small amount of robbing, which would not be harmful when removing supers of liquid honey, can cause great difficulty. Robber bees uncap cells quickly and randomly; in a short period of time, even five minutes, a small number of robbers can ruin the comb surface and appearance of sections even though they may take only a small amount of honey.

Robbing does not occur when a honey flow is in progress. Honey bees prefer natural sources of food, even over honey. The matter is quite different at the end of the flow when they are very alert and will quickly steal any exposed honey. When removing comb honey supers at the end of the honey flow it is best to cover them individually as they are removed.

Bee escapes

In my opinion the best way to remove supers of honey, both comb honey supers and those which are used for liquid honey production, is to use a bee escape. The commercial metal escapes are called "Porter bee escapes" after their inventor. There are also several types of homemade bee escapes which work quite well. The advantages and disadvantages of bee escapes have been discussed by many people in many places;

REMOVING COMB HONEY SUPERS 101

here I am concerned only with their proper use. We have found that most supers will be emptied of bees in 24 hours if the bee escape is properly used.

Bees will not move downward, into the brood nest area or the supers below, if brood is present in the super(s) above the bee escape. Since it is rare that there is brood in comb honey sections one need be little concerned about this aspect of the problem; still, it is well to keep this in mind and to glance at the supers quickly as the escapes are being put into place.

Bees are also reluctant to leave supers if there are bits of broken comb and honey is exposed or dripping from it. Bees will build burr and brace comb between supers which do not have the proper bee space. If one has such broken comb and exposed honey, the only recourse, if escapes are still to be used, is to give the bees a day or so to clean up the honey. When this is done the escapes may be put into place and the bees will then leave the supers within the normal period of time.

Some producers routinely break supers apart (from each other) a few days before they are to be removed from a colony in order to encourage bees to remove any honey from burr and brace comb before the bee escapes are put into place. Whenever supers with excessive burr and brace comb above or below the sections are encountered, they should be marked so that they may be discarded or repaired before the next producing season. Still another reason for breaking supers apart, even when bee escapes are not used, is that honey dripping from broken burr and brace comb may wet supers and sections which are stored under them.

It is helpful to have a special set of innercovers to use as escape boards. Unfortunately it is also expensive. However, if the escape board rims are painted a special color, one is not likely to confuse them with an innercover used for another purpose. It is also helpful to have two Porter bee escapes in an innercover

102 COMB HONEY PRODUCTION

instead of one; the bees will move out of the supers more rapidly and if one escape becomes plugged, as they sometimes do, the bees still have an alternate escape route. Before the escapes are put into place they should be checked to make sure they are not plugged with dead bees or propolis. We have seen escapes propolized by bees, especially in the late fall.

After the escapes have been put into place it is advisable to check the colonies to make certain there are no holes through which robber bees might enter and steal honey. We have found it helpful to carry a small amount of cheap cotton in the truck while removing supers. Grass or leaves can be used to plug holes, crack and crevices, but these wilt and become dislodged easily by robber bees. Cotton is easy to use and stays in place.

There is one great danger in using bee escapes in removing comb honey. The bees cannot control the temperature in a super above an escape board. On occasion one runs the risk of losing a super or more of comb honey because the wax melts. This also makes a great mess. There is additionally a danger of suffocating some bees above the escape board. During warm summer days it is best to use some type of ventilating rim so that the excess heat might escape from the supers. A moving screen works well in this regard. Care must be exercised so that bees cannot enter. Also, if the ventilating rim allows light to enter, the bees in the super will become confused, attracted to the light, and may not be able to escape from the super.

Bee blowers

Bee blowers became popular in the 1960's and are used by a number of commercial beekeepers to remove their honey crop. They work reasonably fast, require only one trip to the apiary, and interestingly, do not make the bees too angry. After an hour or two in the apiary using a bee blower there will be many angry bees in the air, but most of these will have come from colony entrances and are not the bees which are blown from the supers.

When comb honey supers are removed with a bee blower the super is tipped up, perpendicular to the rest of the hive, and the bees blown out of the super and in front of the hive from which the super is being removed. In this way most of them drift back to the correct colony entrance. More time is required to blow bees from a comb honey super than from an extracting super because of the furniture; however, it is a rapid, efficient method for removing comb honey supers. As the supers are removed they should be covered so as to prevent robbing.

Repellents

The two legal[1] bee repellents in the United States are benzaldehyde and butyric anhydride (sold under the trade name *Bee Go*). Both repellents have drawbacks, but will serve reasonably well to remove comb honey. Temperature has a strong effect on repellents: if it is too warm they volatilize too rapidly and may intoxicate and confuse bees; if it is too cool they serve to irritate bees and cause them to uncap cells and eat honey without driving them off the combs.

Fume boards for repellents are rims the dimensions of a standard super, about two inches deep and covered with an absorbent cloth(s) such as burlap. I have found that about four layers of burlap works well. The top (over the burlap) of the repellent board is covered with metal, usually a sheet of galvanized iron. If the iron is painted black it will absorb the sun's rays more rapidly and by being warmer cause the repellent to volatilize quickly.

Using a repellent board successfully requires practice. After the hive cover is removed the bees are smoked lightly; as

---

[1]Carbolic acid and propionic anhydride, both of which will serve to drive bees off combs, were not approved for use by beekeepers by the Environmental Protection Agency at the time of this writing. Carbolic acid is unquestionably the most effective of the known bee repellents.

104 COMB HONEY PRODUCTION

indicated above, smoke is also a repellent and the bees will move away from it. The fume board, which has been wetted with about a tablespoonful of repellent, is placed squarely on the super, but only after the bees have been smoked off the tops of the sections and have moved down the sides of the sections half an inch or so; this may take 5 to 20 seconds. The fume board operator should look under the fume board about 30 seconds after it is first put into place. If the bees are not moving and appear dazed, they are being subjected to too much repellent. If this occurs the repellent board should be removed or placed crossways on the super. Only experience will teach the proper use of fume boards.

The repellents we are now using will not drive bees long distances, usually not more than six or eight inches. For this reason only one comb honey super may be cleared of bees at a time. Usually, however, when the bees have been driven from one super they will have been partially driven from the super below. When one is removing two or more supers at one time it is therefore usually not necessary to smoke the second super to be removed before putting the repellent board on it.

If the comb honey super(s) to be removed is not on the top of the hive I prefer to put it there before using the repellent board. After removing the last filled comb honey super, and especially if there are more comb honey supers still on the colony, I prefer to leave the hive cover off the hive for one or two minutes so that any repellent odor remaining in the hive has time to escape. This cannot be done if bees are robbing. One should not be discouraged if a repellent board does not work perfectly the first time; their successful use requires a great deal of experience.

IX.
Fumigating and Packaging of Comb Honey

There are no chemicals which may be used legally to fumigate comb honey and protect it against attack by wax moths and other insects. Only a few years ago, when the U.S. Department of Agriculture oversaw the use of pesticides, there were several materials which could be used, including cyanide, carbon disulfide, methyl bromide, sulfur and ethylene dibromide.[1] However, when the control of pesticides shifted to the Environmental Protection Agency, a different viewpoint was taken. Elaborate testing procedures are required for each instance in which an insecticide is used. (Any material which kills or repels an insect is an insecticide according to EPA interpretation.)

At the present time, there exists a committee which is examining the so-called use of pesticides on minor crops; this committee's concern is the protection of crops where the quantity of pesticide used is limited and no one company or experiment station can justify the expenditure of funds

---

[1] A common fumigant for combs not included in this list is paradichlorobenzene. This material is useful for protecting combs in storage but may leave a slight residue and flavor in comb honey.

106 COMB HONEY PRODUCTION

required to register a material (pesticide) with the EPA. This committee makes use of information gleaned from the use of a substance on other crops to make a decision relative to its safe use. Since it appears that the use of a fumigant for comb honey may be forthcoming, the writing of this chapter title as it is presented may not be so improper as it first appears.

If comb honey is not consumed immediately, it must be protected from pests. This is not an easy task. While many species of insects may attack and do damage to comb honey, our chief concern is about two of these: the greater wax moth and the lesser wax moth. Methods which will protect comb honey against these two pests will usually serve to protect against the others as well.

Wax moths

The greater wax moth, a primary concern of the comb honey producer, was brought to North America by immigrants from Europe; in fact, wax moths may have been present in the very first hives brought to this country. Like so many of our plant and animal pests, the introduction was accidental and was made at a time when men did not understand that it might be possible to exclude noxious plants and animals through quarantine.

Honey comb is the natural food of the greater wax moth, which is Asian in origin, and is found throughout the world. I have seen wax moths infesting colonies of *Apis dorsata* in the Philippines. European bees are better able to protect themselves against the ravages of wax moths than are the Asian species. Most beekeepers agree that Italian bees are better at ridding a hive of wax moths than are most other races. Beekeepers agree that the best protection against the wax moth is strong, populous colonies.

Wax moths are a great problem in the southern states than in the northern ones because of the long warm season. The greater

FUMIGATING AND PACKAGING OF COMB HONEY

wax moth cannot survive freezing temperatures in any of its life stages: egg, larva, pupa and adult. Small colonies of wax moths overwinter in the North each year in heated cellars and honey houses. Wax moths are used experimentally and are also reared by some people for fish bait (the larvae are used for fish bait). Adults escaping from colonies kept for these purposes, as well as those brought north in the colonies of migrating beekeepers, are the source of our problems with these animals in the North. Because wax moths cannot survive freezing temperatures we recommend that all combs and comb refuse be stored in unheated buildings in the winter in the northern states; in the southern states some form of fumigation is necessary when combs are in storage.

Wax moth eggs are laid in cracks and crevices, often on the outside of a hive, by the fertilized females. The larvae hatch in five to thirty-five days depending upon temperatures; a temperature of 85°F. appears to be optimum for their development. The hatching larvae are very mobile and move into the hive; they are probably attracted by the beeswax odor. If they are not detected by worker bees in the hive, they grow, and may remain in the larval stage for four to five weeks. The rapidity with which the larvae grow depends upon the quality and quantity of the food they receive as well as temperature.

Given a choice, wax moth larvae would seek out old dark comb. Such comb contains pollen, honey and other material which gives them a balanced diet. While wax moths ingest and digest beeswax itself, they must also have vitamins, minerals, etc. as must all animals. Comb honey makes poor food for wax moths. Their growth in sections of comb honey is very slow because there is no pollen, only wax and honey. Still, wax moths ruin a great deal of comb honey every year. They often burrow just under the comb surface, leaving a small amount of webbing and pinpoint size holes in the cappings through which honey may leak. Such perforated cappings "weep", giving the comb surface a poor appearance. In sections which are badly travel stained the growth of the wax moth larvae may be more rapid.

Freezing to protect against moths

Beekeepers who produce only a small quantity of comb honey have found they can protect their sections against the wax moth by freezing them for a short period of time. I am not aware that anyone has studied the precise time the section must be kept in a freezer, but 24 hours should be adequate. Freezing will kill the eggs as well as the other stages in the life cycle of wax moths. A second freezing should be necessary only if one fears adult wax moths may have had access to the sections after a freezing session.

Comb honey stores well in a freezer. The honey retains its delicate flavor. We have kept sections for over a year without their showing any signs of crystallization or deterioration. We usually wrap the sections, but I'm not sure it is necessary. Although we have had no losses of this kind, wax and honey might pick up many strong food odors which might exist in a freezer if they are kept there for a long period of time.

One of the reasons so few men produce comb honey commercially may very well be the problems connected with long-term storage, including the wax moth. It is regrettable, but I advise beekeepers to move their comb honey to market as rapidly as possible because of the several problems concerned with long-term storage. I suspect that men who produce comb honey and cut comb honey in quantity may find the construction of refrigerated rooms helpful since wax moths probably cannot grow at temperatures below 50°F. There is probably no need for lower temperatures for long-term storage. Humidity, which is discussed below, must be watched carefully in such rooms.

Fermentation

Honey is hygroscopic, which means that it will pick up moisture when exposed to the air at normal and high humidities. The moisture content of honey varies greatly. Alfalfa honey from Arizona and southern California, where the humidity is naturally low, may have a moisture content as low as 15 per cent. Mangrove honey from the Florida coast often has a moisture content as high as 21 to 22 per cent. Normally we think of honey as having a moisture content of less than 18 per cent and probably between 17 and 18 per cent is ideal.

Unheated natural honey contains yeast cells. We do not know where these come from. They may be picked up by the bees when they are gathering nectar or they may be part of the natural flora found in a beehive. These yeasts are peculiar in that they will grow in concentrated sugar solutions only. Generally, they grow only in honey containing more than about 19 per cent moisture. When the moisture content is less than 19 per cent, the yeast cells remain dormant and do not grow and cause fermentation. These yeasts produce only a small amount of alcohol and have no commercial value. They are called the osmophilic yeasts because they thrive in concentrated sugar solutions which have a high osmotic pressure; theirs is an unusual environment.

The chief reason that producers of liquid honey pasteurize their product is to kill the yeast cells and to prevent fermentation. Of course, heating has the added value of destroying crystals in the honey and slowing granulation. There is no practical method of pasteurizing comb honey. When fermentation takes place, there is, in addition to a bad odor, the production of carbon dioxide gas. This gas forces some of the honey out of the cells and they "weep", giving the sections a bad appearance.

If comb honey is exposed to high humidity, it will pick up moisture and ferment. The wax cappings over cells are not

33. *A fine section of comb honey with some watery cappings. The darker areas in this section are cells of honey which have absorbed moisture from the air and in which the honey is pushed against the cappings. No damage to the cappings or honey has yet been done in this section. However, if honey in a section such as this continued to absorb moisture, the cappings would be broken and the cells could weep and leak.*

FUMIGATING AND PACKAGING OF COMB HONEY

perfect seals. Water can move through the cappings with ease. It is interesting that only that honey near the surface of the cell may pick up moisture and ferment; the water need not be distributed evenly throughout the honey in a cell for fermentation to take place.

Men who produce large quantities of comb honey usually go to the trouble of storing it in a room with a dehumidifier. The beekeeper with only a few supers of comb honey may remove water from high moisture honey by placing the super in a heated, dry room and blowing air through the supers. Stacking supers of comb honey over an empty super containing a light bulb may be satisfactory but it is a tricky procedure; under the right circumstances the warm air will rise through the supers thus removing moisture-laden air.

There is no way to save fermented comb honey. If the fermentation is only slight, it is sometimes possible to extract the honey from the sections and to pasteurize it. Sometimes, when the fermentation is slight, heating the honey will drive off any objectionable flavor. Badly fermented honey may be fed back to the bees which have the ability to recover and use it.

Cleaning Sections

The cleaning and preparation of comb honey sections for market is slow, hard, tedious work. The propolis must be removed from the wood; this is best done using a knife with a flat, heavy blade. I've seen some beekeepers sand the wood on their sections after they are filled, but this should not be necessary. More propolis is present on sections which have been left on colonies for long periods of time. When one is forced to spend too much time cleaning sections they become aware of the importance of removing sections promptly and before there is an accumulation of propolis.

I've always thought it would be helpful to have an air jet to blow specks (such as smoker ash) from the surface of a finished section before it is wrapped for market. I've used quite successfully a rubber bulb which exhausts a blast of air when it is squeezed. I recommend that one who has a number of sections to clean obtain such a gadget. One can use a fine brush to clean specks off the surface of a comb but sometimes a stiff bristle from such a brush will poke a hole into the wax.

The preparation of sections for market is speeded up if the exposed portions of the wood are coated with paraffin prior to their being placed on a colony, as was suggested in Chapter II.

Packaging

Bee supply manufacturers sell window cartons, and sometimes cellophane wrappers, which may be used as the final package for a section of comb honey. At least one supplier has plastic tubs which may be used for pieces of cut comb honey. I have used all of these items and they work reasonably well; the public is familiar with their design. Since so many beekeepers use the same package, it is important that all beekeepers pack and sell only the best product to maintain the favorable public image which now exists. Beekeepers who use the window cartons usually have a rubber stamp which they use to mark the carton with their name and address.

A major disadvantage of window cartons is that a few people like to squeeze the product they are buying to make certain it is soft and fresh. While they may not break the plastic window, they may rupture the comb surface and honey may leak from the section. I've seen leaky sections on grocery store shelves; the package sticks to the shelf and may attract insects such as ants. The average housewife won't buy such a product and the grocer gets very upset when leakers occur.

FUMIGATING AND PACKAGING OF COMB HONEY 113

Many stores prefer that salesmen inspect and clean the shelf space they use; in the case of comb honey this is probably a good idea.

The only defense against leaking comb honey sections is to place the sections inside plastic bags in the window carton. This is an added expense and nuisance but in most markets is probably necessary to be successful. The inner plastic bag should be selected with care so that its appearance doesn't destroy the image of the section through the window of the outer package.

Sun bleaching comb honey sections

Comb honey producers of decades ago were extremely fussy about the appearance of their comb honey sections. They wanted sections with cappings as white and light as possible. Many of them, in addition to selecting bees which made white cappings, would expose their finished comb honey sections to sunlight to bleach them further. As late as 1950 there was one comb honey producer in northern New York who had a greenhouse-like structure on the southern side of his shop. The room was approximately 20 feet long. Along the glass side were shelves five inches high which would hold comb honey sections, one deep. The sections would be exposed to sunlight for several days, the face toward the sunlight being rotated after two to four days.

Even today producers of highly refined beeswax for the dental trade sun bleach their wax. Exposure of beeswax to direct sunlight for five to seven days will do much to lighten its color. Of course, in the case of bleaching sections, one must exercise great care that they not be kept in a room which will become overheated causing the wax to melt. I suggest that sun bleaching is a refinement not necessary in today's market.

114 COMB HONEY PRODUCTION

> ### Basswood Honey
>
> The American basswood tree, *Tilia americana*, is the predominant source of this honey. In the Ithaca area the production of nectar by flowering basswood trees is one of the most predictable honey flows—weather permitting, it starts on July 6, 7 or 8. However, the honey flow, sometimes called the nectar flow, is erratic; a dry summer or too much rain can have an adverse effect on nectar secretion.
>
> The flavor of basswood honey is unique and easy to distinguish. Most honey packers avoid adding basswood honey to their blends; they feel the sharp, minty flavor is too strong for the average American palate. People who enjoy tasting varietal honeys will find basswood honey a special treat.

34. A second label, two by three inches, which may be used on a jar of liquid honey or a section of comb honey. The thought behind the label is to tell the customer more about the product; hopefully this might aid in sales. Second labels are sometimes called back labels.

Second labels (back labels)

A back label is a second label on a package, which describes more fully the contents of the package. They are used successfully by a number of manufacturers and can add much to a section or jar of honey. For example, a beekeeper might write 50 or so words about a special kind of honey. It may be helpful to write about comb honey itself, stating that the honey bee secretes wax which it then molds into place to make the honey

FUMIGATING AND PACKAGING OF COMB HONEY

comb. Second labels may also be used to describe a special soil type or area where the honey is produced. One might write about the fact that beeswax may be chewed and swallowed without difficulty; not everyone understands that fact.

Comb honey is a luxury product and people who buy it want it for a special reason. The more they can be told about what they are buying and how it was produced, the more likely they are to buy again. While it is true that much of the American public buys prepared, pre-cooked food, there will always be a market for some gourmet items. The producer of comb honey can take advantage of his special knowledge to encourage this market. Even when comb honey is produced and packaged as a gift for one's friends, the added information can be helpful.

Marketing experts agree that certain figures and words should be avoided when writing or preparing labels. Some people find a picture of a bee offensive; they tend to think of all insects as being the same thing. A characterization of a bee may be satisfactory. It is probably better to talk about a honey sac than a honey stomach. One can say, as I did above, that bees mold wax into place in the comb, rather than that they chew the wax prior to working it into the comb.

Shipping comb honey

One might write only a single word under this heading; that would be "don't". The bee journals are full of tales of grief and laments by those who have shipped comb honey by the case or in individual packages. Most beekeepers who ship honey by mail send liquid or crystallized honey in metal or plastic containers only. Comb honey does not ship well; the comb may break from the wood and/or sections may leak. The comb honey producer who mails his produce is forced to package the honey so well that the cost is prohibitive.

Weight laws

States vary as regards laws about marking the weight of packaged food products on the package. In New York State, for example, it is not necessary to indicate the weight of the honey in a package of comb honey. The weight of the honey in a glass jar must be indicated, but in New York a section of comb honey is sold as a unit. However, many states require that the net weight be indicated on the package. Most window cartons sold by bee supply companies have a line on the carton which reads "Net wt. (blank to be filled in) oz." In some states, the producer who sells his honey must fill this in and in others he need not do so. It is necessary to check with a local authority or the State Apiary Inspector or Extension Apiculturist to determine what is required in each state.

X.
Showing and Grading Comb Honey

State and county fairs are becoming increasingly popular in the United States as more attention is being paid to rural living. At the same time, the honey shows put on by such organizations as the Eastern Apicultural Society and the Amercan Beekeeping Federation are also serving a useful function by calling the public's attention to high quality products.

Many people seem reluctant to exhibit in competition for fear of not doing well or arousing a judge's ire. I don't think a beekeeper should be concerned about either of these items. There is no better way to learn some of the do's and don't's concerning both the production and packaging of comb honey than to enter into competition. Of course, there is no greater satisfaction than carrying home a silver bowl and a blue ribbon. I have encouraged greater participation in honey shows and shall continue to do so.

Official grades for comb honey

The Standardization Section of the Processed Products

Standardization and Inspection Branch of the Fruit and Vegetable Division of the Agricultural Marketing Service of the United States Department of Agriculture, Washington, D.C., is responsible for establishing grades for comb honey. The present grades are contained in an 8-page brochure entitled *United States Standards for Grades of Comb Honey*. These were effective May 24, 1967 and supersede grades which have been in effect since August 1933. Single copies of these grades are available at no charge by writing the Standardization Section at the above address.

Throughout the brochure on grading, mention is made of standard color charts. Unfortunately, these are not available and copies are rare.

While it is correct that the American consuming public seems to prefer lighter, milder honeys, color should not be the primary consideration in grading honey. There are some very delightful, dark, strong-flavored comb honeys which are much preferred by some people. Of course, it is well to differentiate between sections of comb honey which are dark because of the honey they contain and those which are dark because of travel stain.

At the present time the color of liquid honey is officially established by the U.S. Department of Agriculture glass color standards. On an official basis these are used in place of the Pfund grader which is widely used in the industry. The color grader recommended by the USDA, sometimes called the color comparator, has the normal fault of such items in that in time there is some fading of the color materials, depending upon exposure to light etc. Also, the liquids in the jars may evaporate, making them useless. I have no doubt that someday the Pfund grader will again become the official standard by which liquid honey is graded. The USDA color grader was adopted by that group in 1951 upon recommendations by persons not thoroughly familiar with the industry and honey.

The glass color standards are available from a supplier licensed by the USDA:

Phoenix Precision Instrument Division
The Virtis Company, Inc.
Route 208
Gardiner, New York 12525

Outlined below are the grades for U.S. Fancy[1] comb section honey. Reprinting all of the grades would take too much space. The purpose in reprinting this section is to point out the factors which judges consider when judging comb honey.

"U.S. Fancy honey shall consist of comb-section honey that meets the following requirements:

(a) The comb shall —

(1) Have no uncapped cells except in the row attached to the wood section;

(2) Be attached to 75 percent of the adjacent area of the wood section if the outside row of cells is empty, or attached to 50 percent if the outside row is filled with honey;

(3) Not project beyond the edge of the wood section;

(4) Not have dry holes;

(5) Have not more than a total of 2 1/2 linear inches of through holes;

(6) Be free from cells of pollen.

(b) The cappings shall —

(1) Be dry and free from weeping and from damage caused by bruising or other means.

---

[1]Comb honey is graded as follows: U.S. Fancy, U.S. No. 1, U.S. No. 2, and "Unclassified." Unclassified is that "which does not conform to the requirements for any of the foregoing grades." Interestingly, extracted (liquid) honey is graded as Grade A, B, C, or D. Grades for extracted honey are contained in a six-page brochure entitled "United States Standards for grades of Extracted Honey, effective April 16, 1951". This brochure is available from the same agency which offers the standards for comb honey.

(2) Present a uniformly even appearance except in the row attached to the wood section.

(c) The color of the comb and cappings shall conform to the requirements as illustrated for this grade in the official color chart.

(d) The honey shall —

(1) Be uniform in color throughout the comb.

(2) Be free from damage caused by granulation, honeydew, poorly ripened or sour honey, objectionable flavor or odor, or other means.

(e) The wood section shall —

(1) Be as free from excessive propolis and/or pronounced stains as illustration A in the official color chart.

(2) Be smooth and new in appearance, of white to light buff basswood, and shall not contain knots and/or streaks in excess of the amount shown in illustration B in the color chart.

(f) The minimum net weight shall be 12 ounces, unless otherwise specified."

Judging comb honey

Judging honey is not an easy task. Many committees, notably those formed by the American Beekeeping Federation, have attempted to make uniform rules for judging. The chief problem is that judges differ as to what they think is and is not important. I freely admit that I have my prejudices too in this regard. I think it is important to place emphasis on those factors which the beekeeper can best control. For example, most beekeepers find it difficult to control the moisture content of their honey. Honeys produced in areas where the humidity is low will be low in moisture; the reverse is likewise true. While it is possible to remove moisture from honey in the comb using dehumidifiers or special ventilating devices, most beekeepers do not have such equipment. Thus, in my opinion, a judge should consider moisture content, but he should not give the same importance to moisture content that he assigns to travel stain, uncapped cells, clean wood, etc.

The judging cards used by the Eastern Apicultural Society for comb and chunk honey classes in their annual honey show are illustrated. It is worthwhile to consider each of these questions; no doubt some persons would care to assign the points differently but this is really a minor consideration.

Figure 35: Comb Honey and Bulk Honey Frame:
Uniformity of appearance — In the case of comb honey the entrant is required to submit three sections while in the case of a bulk honey frame only one frame is required. In the case of sections, they should look alike; this includes the opposite sides. The honey in each section should look and taste alike. It is safest to take the sections from the same hive in the case of sections, or at least from the same apiary. In both cases the comb surface should be even and without indentations or areas which protrude.

Absence of uncapped cells — Only rarely does one find a section or a comb without a few uncapped cells; however, this does occur frequently with the round Cobana sections and is one of the reasons the Cobanas are preferred by some producers. Uncapped cells adjacent to the wood are less serious than are those which are one or two cells removed from the edge of the section. In some contests judges may be forced to count the number of uncapped cells.

Uncapped cells are also those capped cells which have been perforated by bees when the honey is being removed from the hive. Bees which are smoked will engorge with honey. If there are no open cells where they happen to be standing when smoked, they will quickly uncap a small area, but one large

122 COMB HONEY PRODUCTION

EASTERN APICULTURAL SOCIETY
JUDGE'S SCORE CARD

Event: COMB HONEY & BULK HONEY FRAME   Class: _____   Entry No.: ___

Point Scoring	Item	Judge's Remarks
20	Uniformity of appearance	
10	Absence of uncapped cells	
15	Uniformity of color	
10	Absence of watery cappings	
15	Cleanliness of section or frame	
5	Freedom from granulation & pollen	
15	Uniformity of weight	
10	Total weight of entry	
100		Award: _____

35-36. Typical competition score cards. While not every judge would agree with the distribution of score points, these cards make reference to the important considerations made by judges as they inspect and score sections and combs of honey.

EASTERN APICULTURAL SOCIETY   JUDGE'S SCORE CARD
Event: CHUNK HONEY   ·Class: _____   Entry No.: _____

Point Scoring	Item	Judge's Remarks
20	Neatness of cut Ragged edges, parallel cuts, four-sided cut, & uniformity of size of cut	
20	Absence of watery cappings, uncapped cells and pollen cells	
20	Cleanliness No travel stain, specks of foreign matter, flakes of wax, foam and crystallization	
30	Uniformity of appearance Uniformity of capping structure, color and thickness of comb	
10	Density and flavor of liquid part	
100		Award: _____

SHOWING AND GRADING COMB HONEY

enough to insert their proboscis so as to take up honey. Robber bees will also perforate cappings quickly and ruin the appearance of a section of comb. Exposing the combs only five or ten minutes may be sufficient for robber bees to make many such holes and to harm many sections.

Uniformity of color — There is little problem with uniformity of color during a rapid honey flow. However, when the flow takes place over a long period of time, it is possible that bees may gather nectar from many sources and the nectars may not be of the same color. Nothing stands out so boldly as a single cell of dark honey in a light colored comb honey section, or vice versa. Lack of uniformity in color is a serious matter, especially to the consuming public which is increasingly accustomed to uniformity in the food products they buy; it is for this reason that this factor is assigned more points.

Absence of watery cappings — Bees normally build the cell cappings in such a way that they do not touch or press firmly against the honey in the cells. However, honey is hygroscopic, which means it picks up water easily. Often a honey will pick up moisture, say moving from a moisture content of 17.5 per cent to 18.0 or 18.5 per cent in a matter of two or more days in a room with high humidity. In most areas the humidity fluctuates widely and it is for this reason that it is recommended that honey be stored in a warm, dry place. When honey picks up moisture its volume swells and it may push against the cappings, giving them a darker, watery appearance. If the honey picks up too much moisture, fermentation may take place on the surface of the honey in the cell, just under the cappings. When this occurs the sections leak honey from the cells and their appearance is ruined. Watery cappings with honey spilling our of the cells are a sign that fermentation is about to occur and thus the importance of this part of the score card.

Cleanliness of the section or frame — Once in a great while a show will have a class for a frame of honey to be extracted. In such a case the wood may be old and dark in color; however, it should be uniformly dark and without cracks, chips or breaks. In most shows the judge requires that bulk honey frames, as well as sections, be in new white wood.

Freedom from granulation and pollen — Granulation is undesirable for two reasons: honey with coarse crystals does not have as fine a flavor as does liquid honey; perhaps more important, honey which is partially crystallized may ferment. When honey granulates, only a small part of the water in the honey is taken into the sugar crystals. Thus, the moisture content of that which is left is higher than it was before granulation started. Yeasts are present in all honey and in liquid honey are usually killed by pasteurization. There is no practical way to prevent granulation of honey in the comb. When the moisture content of an unpasteurized honey gets above about 19 per cent it will ferment. Some honeys tend to granulate more rapidly than do others but all honeys granulate most rapidly at 57°F, plus or minus about ten degrees. When a show is in the winter, sometime after the producing season, we have found it satisfactory to store the sections or combs in a freezer. This both prevents them from granulating and from adding moisture.

Pollen has a bitter taste and a cell of pollen in a section or comb of honey stands out like the leaves of a sugar maple in a pine forest in the fall. Judges are quick to penalize such a fault. Keeping cells in a comb honey section free of pollen has been discussed; it is not always easy to do.

Uniformity of weight — This consideration has no application to a bulk honey frame. Bulk honey frames which are thicker at one end than the other are usually discounted under the uniformity of appearance consideration. Insofar as comb honey sections are concerned they should be of equal weight. I am often asked why a judge or a show requests three samples in liquid and comb honey classes. The reason is simply to make certain that the exhibitor can produce more than one perfect section or jar of honey. Uniformity in the product to be marketed is important. I, and I think most judges, are aware that it is really the consuming public which is the judge of our product. Whenever there is a question about one of the judging considerations it is well to try to place oneself in the place of a buyer of a section or jar of honey.

Total weight of entry — Some states have a net weight law and the weight of the section must be stamped on the package. In other states the section is sold as a unit and the weight is not important. Still, a section should be reasonably full, weighing at least 12 to 13 ounces. In most shows sections receive the full number of points when weight is considered.

Figure 36: Chunk honey:
Neatness of cut etc. — The piece of chunk honey in a jar of liquid honey should be rectangular or square, depending upon the shape of the jar. Most pieces of chunk honey usually rate fairly high in this category and it is only a question of paying attention to detail. It is unfortunate when an exhibitor's cutting and care of a piece of chunk honey do not match the expertise of the bees.

Absence of watery cappings etc. — These considerations have been discussed above. It is interesting that watery cappings are as visible when the comb is placed in a jar of liquid honey as they are in an exposed section. Also, honey in the section does not, under normal circumstances, pick up moisture from the liquid honey which surrounds it.

Cleanliness etc. — All the considerations listed are easily noted by a judge. Additionally, there must not be pieces of wax floating in the liquid honey which surrounds the piece(s) of chunk honey. It is important to check the jars just before the show to make certain no loose pieces of wax have broken from the comb.

Foam in any liquid honey pack is to be avoided. Air enters honey when it is extracted, pumped or poured. These air bubbles must be allowed to settle to the surface before the honey is packed. If I am critical of any one thing about honey packed by beekeepers, it is that all too often there is too much foam in the form of a ring of white froth around the top inside of the jar. Foam is not present in the jars of honey packaged by large, commercial honey packers and the consuming public does not understand why it would be present in that packed by the producer-packer. It is true that it is impossible to avoid incorporating air into the honey when it is extracted; in other instances the beekeeper can exercise care which might otherwise cause problems. The only way to remove air from liquid honey it to give it time to settle, rise to the surface and be skimmed off in the packing tank.

Density and flavor of liquid part — The moisture content of the liquid honey in a chunk honey pack is measured and considered in the same way one considers the honey in the liquid classes. Jars which contain honey with more than 18.6 per cent moisture are disqualified.

The flavor of a honey is always the most debatable part of any show. If a honey has a burned or fermented flavor it is easy enough to discount it. However, to judge beyond that is difficult. In my opinion a honey which tastes of privet, or something similar, should be discounted. (Privet produces a honey which is dark and has a bitter taste; fortunately such honeys are rare.) Opinions vary as to how dark the honey in a liquid or chunk honey pack might be. A small number of consumers prefer a dark, strong tasting honey; others think that such honeys should be sold in the bakery trade only. Judges need to be wary when they trespass on the honey taste and color preferences of beekeepers.

Index

alarm odor, 61
American Beekeeping Federation, 117, 120
American foulbrood, 53-54, 57
apiary sites, 46-51

back labels (see second labels)
bait sections, 70, 93-94
basswood lumber, 10, 43
bee blowers, 102-103
bee escapes, 100-102
Bee Go, 103
benzaldehyde, 103
bleaching comb honey, 113
bottomboards, 15-17, 21-23, 50-51
brace comb, 52, 101
brood patterns, 62, 69
burr comb, 29, 52, 62, 101
butyric anhydride, 103

carbolic acid, 103
chimney effect, 91-92
Churchill, Raymond, 77-80
cleaning frames, 85
cleaning sections, 111
clipping queens, 62
closing upper entrances, 92
cobana sections, 32-35
Coggshall, Millard, 83
comb honey
   adulteration, 11, 13
   bleaching sections, 113
   cleaning sections, 111-112
   cobana sections, 32-35
   definition, 9-10
   fermentation, 111
   foundation, 36, 45
   grades for, 117, 121
   judging, 120-126
   labeling, 114-115
   pollen in, 19
   sections, 27, 45
   shipping, 115
   showing, 117-126
   split sections, 29-30
   supers, 27, 45
   travel stain, 14
   watery cappings, 110, 123, 125

weeping, 107-108
weight laws, 116
combination cover-bottomboards, 20
congestion, 66
covers, 20-21
cut comb honey, 81-87

Demuth, George, 15, 90
dimensions for equipment, 17
division board feeders, 56
dry sugar, 56
dummy boards (see follower boards)

Eastern Apicultural Society, 77, 117, 121-122
8-frame hive, 24
equipment
   bee escapes, 100-102
   bottomboards, 15-17, 21-23
   combination cover-bottomboards, 20
   covers, 20-21
   excluders, 18, 45, 83
   follower boards, 20
   frames, 18-19, 84-86
   hives, 8 and 10-frame, 24-26
   hivestands, 21-23
   innercovers, 20-21
   metal eyelets, 19
   repellent boards, 104
   sections, 27-45, 56-57, 111-112
   supers, 27-45, 78-79
European foulbrood, 57
examining colonies, 58-61
excluders, 18, 45, 83
eyelets, metal, 19

fastening foundation, 39
feeding colonies, 55-57, 95-97
feeding dry sugar, 56
fermentation, 109-111
foam on honey, 126
follower boards, 20
food chamber, 75
foundation, 35-43, 67, 84-85, 88-89
frames, 18-19, 84-86
freezing comb honey, 108
fume boards, 103
fumigating comb honey, 105-106

128 INDEX

glass color standards, 119
go-backs, 94-97
grades for comb honey, 117
granulation, 124
greater wax moth, 106

Harbison, J. S., 10
hives, 8 and 10-frame, 24-26
hivestands, 21-23
honey
  adulteration, 11-15
  alfalfa, 46-74
  basswood, 47, 52, 63
  buckwheat, 46
  clover, 52, 74
  color, 46
  flavor, 46
  foam on, 126
  goldenrod, 46, 52, 74-75
  honeydew, 46
  mangrove, 109
  privet, 47
  wild thyme, 46
Honey Industry Council, 12
humidity, effect on comb honey, 109
hygroscopicity, 109-111

innercovers, 20-21
Italian bees, 106

judging comb honey, 120

keeping bees in congested areas, 48
Killion, Carl, 15, 25, 31, 43
Kruse, Charles A., 25

Langstroth, L. L., 24
leveling colonies, 50, 90-92
lie, Wiley, 11
lye, 31, 85

metal eyelets, 19
Miller, C. C. (Dr.), 15, 25, 72-73, 77
moldy comb, 54
mounting blocks, 39

net weight laws, 116
nosema, 57

opening colonies, 58-61
outbuildings, 51

packaging comb honey, 86, 112
packaging cut comb honey, 86
packing bees (see wintering)
paradichlorobenzene, 105
Pellett, Frank, 10
pentachlorophenol, 17, 22
Phillips, E. F., 24

pollen, 19, 83, 124
pollen, taste of, 9-10, 19
Porter bee escapes, 100
preparation for the honey flow, 67
propionic anhydride, 103
propolis, 14, 18-19, 52, 62, 88, 111
pure food and drug laws, 11-12

queen excluders, 18, 45, 83

races of bees, 76
repellents, 68, 103-104
reversing supers, 78
robbing, 100
Root, A. I., 11
Root, E. R., 24

sacbrood, 57
Schraeder, Clarence, 25, 31
score cards, 122
seasonal management, 52-80, 82-83
second labels, 114-115
section press, 40
sections, 27-45, 56-57, 111-112
Seeley, Thomas, 26
shade for bees, 49-50
shipping comb honey, 115
smoker ash, 99-100, 112
smoking bees, 58-61, 99-100
split sections, 29-30
spreading split sections, 42
starter sections (see bait sections)
Stevens, Gerald, 25-26
supering, 88-97
supers, 27-45, 78-79
swarm control, 65-66, 70-73
swarm prevention, 63, 65-66, 79-80
swarming season, 65

"T" super, 30-31
travel stain, 14, 67, 89, 98
turning comb honey supers, 92

upper entrances, 92-93
U.S. Department of Agriculture, 11-12, 105, 118-120

virgin honey, 9

water for bees, 48-49
watery cappings, 110, 125
wax moths, 106-108
weight laws, 116
weeping, 107
Wiley lie, 11
windbreaks, 49
wintering bees, 53, 76, 80
wood preservative, 17, 22

www.ingramcontent.com/pod-product-compliance
Lightning Source LLC
Chambersburg PA
CBHW051543230426
43669CB00015B/2709